Iterative Methods for Solving
Linear Systems

Frontiers in Applied Mathematics

Frontiers in Applied Mathematics is a series that presents new mathematical or computational approaches to significant scientific problems. Beginning with Volume 4, the series reflects a change in both philosophy and format. Each volume focuses on a broad application of general interest to applied mathematicians as well as engineers and other scientists.

This unique series will advance the development of applied mathematics through the rapid publication of short, inexpensive books that lie on the cutting edge of research.

Frontiers in Applied Mathematics

Iterative Methods for Solving Linear Systems

Anne Greenbaum
University of Washington
Seattle, Washington

Society for Industrial and Applied Mathematics
siam.
Philadelphia 1997

Library of Congress Cataloging-in-Publication Data

Greenbaum, Anne.
    Iterative methods for solving linear systems / Anne Greenbaum.
        p.   cm. -- (Frontiers in applied mathematics ; 17)
    Includes bibliographical references and index.
    ISBN 0-89871-396-X (pbk.)
    1. Iterative methods (Mathematics)   2. Equations, Simultaneous-
-Numerical solutions. I. Title. II. Series.
QA297.8.G74   1997
519.4--dc21   97-23271

siam is a registered trademark.

Contents

List of Algorithms

Preface

In recent years much research has focused on the efficient solution of large sparse or structured linear systems using iterative methods. A language full of acronyms for a thousand different algorithms has developed, and it is often difficult for the nonspecialist (or sometimes even the specialist) to identify the basic principles involved. With this book, I hope to discuss a few of the most useful algorithms and the mathematical principles behind their derivation and analysis. The book does not constitute a complete survey. Instead I have tried to include the most *useful* algorithms from a practical point of view and the most *interesting* analysis from both a practical and mathematical point of view.

The material should be accessible to anyone with graduate-level knowledge of linear algebra and some experience with numerical computing. The relevant linear algebra concepts are reviewed in a separate section and are restated as they are used, but it is expected that the reader will already be familiar with most of this material. In particular, it may be appropriate to review the QR decomposition using the modified Gram–Schmidt algorithm or Givens rotations, since these form the basis for a number of algorithms described here.

Part I of the book, entitled *Krylov Subspace Approximations*, deals with general linear systems, although it is noted that the methods described are most often useful for very large sparse or structured matrices, for which direct methods are too costly in terms of computer time and/or storage. No specific applications are mentioned there. Part II of the book deals with *Preconditioners*, and here applications must be described in order to define and analyze some of the most efficient preconditioners, e.g., multigrid methods. It is assumed that the reader is acquainted with the concept of finite difference approximations, but no detailed knowledge of finite difference or finite element methods is assumed. This means that the analysis of preconditioners must generally be limited to model problems, but, in most cases, the proof techniques carry over easily to more general equations. It is appropriate to separate the study of iterative methods into these two parts because, as the reader will see, the tools of analysis for Krylov space methods and for preconditioners are really quite different. The field of preconditioners is a much broader one, since the derivation of the preconditioner can rely on knowledge of an underlying problem from which the linear system arose.

This book arose out of a one-semester graduate seminar in Iterative Methods for Solving Linear Systems that I taught at Cornell University during the fall of 1994. When iterative methods are covered as part of a broader course on numerical linear algebra or numerical solution of partial differential equations, I usually cover the overview in section 1.1, sections 2.1–2.4 and 3.1, and some material from Chapter 5 and from Part II on preconditioners.

The book has a number of features that may be different from other books on this subject. I hope that these may attract the interest of graduate students (since a number of interesting open problems are discussed), of mathematicians from other fields (since I have attempted to relate the problems that arise in analyzing iterative methods to those that arise in other areas of mathematics), and also of specialists in this field. These features include:

- A brief overview of the state of the art in section 1.1. This gives the reader an understanding of what has been accomplished and what open problems remain in this field, without going into the details of any particular algorithm.

- Analysis of the effect of rounding errors on the convergence rate of the conjugate gradient method in Chapter 4 and discussion of how this problem relates to some other areas of mathematics. In particular, the analysis is presented as a matrix completion result or as a result about orthogonal polynomials.

- Discussion of open problems involving error bounds for GMRES in section 3.2, along with exercises in which some recently proved results are derived (with many hints included).

- Discussion of the transport equation as an example problem in section 9.2. This important equation has received far less attention from numerical analysts than the more commonly studied diffusion equation of section 9.1, yet it serves to illustrate many of the principles of non-Hermitian matrix iterations.

- Inclusion of multigrid methods in the part of the book on preconditioners (Chapter 12). Multigrid methods have proved extremely effective for solving the linear systems that arise from differential equations, and they should not be omitted from a book on iterative methods. Other recent books on iterative methods have also included this topic; see, e.g., [77].

- A small set of recommended algorithms and implementations. These are enclosed in boxes throughout the text.

This last item should prove helpful to those interested in solving particular problems as well as those more interested in general properties of iterative methods. Most of these algorithms have been implemented in the *Templates for the Solution of Linear Systems: Building Blocks for Iterative Methods* [10], and the reader is encouraged to experiment with these or other iterative routines for solving linear systems. This book could serve as a supplement to the *Templates* documentation, providing a deeper look at the theory behind these algorithms.

I would like to thank the graduate students and faculty at Cornell University who attended my seminar on iterative methods during the fall of 1994 for their many helpful questions and comments. I also wish to thank a number of people who read through earlier drafts or sections of this manuscript and made important suggestions for improvement. This list includes Michele Benzi, Jim Ferguson, Martin Gutknecht, Paul Holloway, Zdeněk Strakoš, and Nick Trefethen.

Finally, I wish to thank the Courant Institute for providing me the opportunity for many years of uninterrupted research, without which this book might not have developed. I look forward to further work at the University of Washington, where I have recently joined the Mathematics Department.

Anne Greenbaum
Seattle

Chapter 1

Introduction

The subject of this book is what at first appears to be a very simple problem—how to solve the system of linear equations $Ax = b$, where A is an n-by-n nonsingular matrix and b is a given n-vector. One well-known method is Gaussian elimination. In general, this requires storage of all n^2 entries of the matrix A as well as approximately $2n^3/3$ arithmetic operations (additions, subtractions, multiplications, and divisions). Matrices that arise in practice, however, often have special properties that only partially can be exploited by Gaussian elimination. For example, the matrices that arise from differencing partial differential equations are often *sparse*, having only a few nonzeros per row. If the (i, j)-entry of matrix A is zero whenever $|i - j| > m$, then a *banded* version of Gaussian elimination can solve the linear system by storing only the approximately $2mn$ entries inside the band (i.e., those with $|i - j| \leq m$) and performing about $2m^2n$ operations. The algorithm cannot take advantage of any zeros inside the band, however, as these *fill in* with nonzeros during the process of elimination.

In contrast, sparseness and other special properties of matrices can often be used to great advantage in matrix–vector multiplication. If a matrix has just a few nonzeros per row, then the number of operations required to compute the product of that matrix with a given vector is just a few n, instead of the $2n^2$ operations required for a general dense matrix–vector multiplication. The storage required is only that for the nonzeros of the matrix, and, if these are sufficiently simple to compute, even this can be avoided. For certain special dense matrices, a matrix–vector product can also be computed with just $O(n)$ operations. For example, a Cauchy matrix is one whose (i, j)-entry is $1/(z_i - z_j)$ for $i \neq j$, where $z_1, \ldots, z_n$ are some complex numbers. The product of this matrix with a given vector can be computed in time $O(n)$ using the fast multipole method [73], and the actual matrix entries need never be computed or stored. This leads one to ask whether the system of linear equations $Ax = b$ can be solved (or an acceptably good approximate solution obtained) using matrix–vector multiplications. If this can be accomplished with a moderate number of matrix–vector multiplications and little additional work, then the iterative procedure that does this may far outperform Gaussian elimination in terms of both work and storage.

One might have an initial guess for the solution, but if this is not the case, then the only vector associated with the problem is the right-hand side vector b. Without computing any matrix–vector products, it seems natural to take some multiple of b as the first approximation to the solution:

$$x_1 \in \operatorname{span}\{b\}.$$

One then computes the product Ab and takes the next approximation to be some linear combination of b and Ab:

$$x_2 \in \operatorname{span}\{b, Ab\}.$$

This process continues so that the approximation at step k satisfies

(1.1) $$x_k \in \operatorname{span}\{b, Ab, \ldots, A^{k-1}b\}, \quad k = 1, 2, \ldots.$$

The space represented on the right in (1.1) is called a *Krylov subspace* for the matrix A and initial vector b.

Given that x_k is to be taken from the Krylov space in (1.1), one must ask the following two questions:

(i) How good an approximate solution is contained in the space (1.1)?

(ii) How can a good (optimal) approximation from this space be computed with a moderate amount of work and storage?

These questions are the subject of Part I of this book.

If it turns out that the space (1.1) does not contain a good approximate solution for any moderate size k or if such an approximate solution cannot be computed easily, then one might consider modifying the original problem to obtain a better Krylov subspace. For example, one might use a *preconditioner* M and effectively solve the modified problem

$$M^{-1}Ax = M^{-1}b$$

by generating approximate solutions $x_1, x_2, \ldots$ satisfying

(1.2) $$x_k \in \operatorname{span}\{M^{-1}b, (M^{-1}A)M^{-1}b, \ldots, (M^{-1}A)^{k-1}M^{-1}b\}.$$

At each step of the preconditioned algorithm, it is necessary to compute the product of M^{-1} with a vector or, equivalently, to solve a linear system with coefficient matrix M, so M should be chosen so that such linear systems are much easier to solve than the original problem. The subject of finding good preconditioners is a very broad one on which much research has focused in recent years, most of it designed for specific classes of problems (e.g., linear systems arising from finite element or finite difference approximations for elliptic partial differential equations). Part II of this book deals with the topic of preconditioners.

1.1. Brief Overview of the State of the Art.

In dealing with questions (i) and (ii), one must consider two types of matrices—
Hermitian and *non-Hermitian*. These questions are essentially solved for the
Hermitian case but remain wide open in the case of non-Hermitian matrices.

A caveat here is that we are now referring to the *preconditioned* matrix. If
A is a Hermitian matrix and the preconditioner M is Hermitian and *positive
definite*, then instead of using left preconditioning as described above one can
(implicitly) solve the modified linear system

$$(1.3) \qquad\qquad L^{-1}AL^{-H}y = L^{-1}b, \quad x = L^{-H}y,$$

where $M = LL^H$ and the superscript ^{H} denotes the complex conjugate
transpose ($L_{ij}^H = \bar{L}_{ji}$). Of course, as before, one does not actually form the
matrix $L^{-1}AL^{-H}$, but approximations to the solution y are considered to
come from the Krylov space based on $L^{-1}AL^{-H}$ and $L^{-1}b$, so the computed
approximations to x come from the space in (1.2). If the preconditioner M
is *indefinite*, then the preconditioned problem cannot be cast in the form
(1.3) and treated as a Hermitian problem. In this case, methods for non-
Hermitian problems may be needed. The subject of special methods for
Hermitian problems with Hermitian indefinite preconditioners is an area of
current research.

Throughout the remainder of this section, we will let A and b denote the
already preconditioned matrix and right-hand side. The matrix A is Hermitian
if the original problem is Hermitian and the preconditioner is Hermitian and
positive definite; otherwise, it is non-Hermitian.

1.1.1. Hermitian Matrices.

For *real symmetric* or *complex Hermitian*
matrices A, there is a known short recurrence for finding the "optimal" approximation of the form (1.1), if "optimal" is taken to mean the approximation
whose residual, $b - Ax_k$, has the smallest Euclidean norm. An algorithm that
generates this optimal approximation is called MINRES (minimal residual)
[111]. If A is also positive definite, then one might instead minimize the A-norm
of the error, $\|e_k\|_A \equiv \langle A^{-1}b - x_k, b - Ax_k\rangle^{1/2}$. The conjugate gradient (CG)
algorithm [79] generates this approximation. For each of these algorithms, the
work required at each iteration is the computation of one matrix–vector product (which is always required to generate the next vector in the Krylov space)
plus a few vector inner products, and only a few vectors need be stored. Since
these methods find the "optimal" approximation with little extra work and
storage beyond that required for the matrix–vector multiplication, they are
almost always the methods of choice for Hermitian problems. (Of course, one
can never really make such a blanket statement about numerical methods. On
some parallel machines, for instance, inner products are very expensive, and
methods that avoid inner products, even if they generate nonoptimal approximations, may be preferred.)

Additionally, we can describe precisely how good these optimal approximations are (for the worst possible right-hand side vector b) in terms of the *eigenvalues* of the matrix. Consider the 2-norm of the residual in the MINRES algorithm. It follows from (1.1) that the residual $r_k \equiv b - Ax_k$ can be written in the form

$$r_k = P_k(A)b,$$

where P_k is a certain kth-degree polynomial with value 1 at the origin, and, for any other such polynomial p_k, we have

(1.4) $$\|r_k\| \leq \|p_k(A)b\|.$$

Writing the eigendecomposition of A as $A = Q\Lambda Q^H$, where $\Lambda = \text{diag}(\lambda_1, \ldots, \lambda_n)$ is a diagonal matrix of eigenvalues and Q is a unitary matrix whose columns are eigenvectors, expression (1.4) implies that

$$\|r_k\| \leq \|Qp_k(\Lambda)Q^H b\| \leq \|p_k(\Lambda)\| \, \|b\|,$$

and since this holds for any kth-degree polynomial p_k with $p_k(0) = 1$, we have

(1.5) $$\|r_k\|/\|b\| \leq \min_{p_k} \max_{i=1,\ldots,n} |p_k(\lambda_i)|.$$

It turns out that the bound (1.5) on the size of the MINRES residual at step k is *sharp*—that is, for each k there is a vector b for which this bound will be attained [63, 68, 85]. Thus the question of the size of the MINRES residual at step k is reduced to a problem in approximation theory—how well can one approximate zero on the set of eigenvalues of A using a kth-degree polynomial with value 1 at the origin? One can answer this question precisely with a complicated expression involving all of the eigenvalues of A, or one can give simple bounds in terms of just a few of the eigenvalues of A. The important point is that the norm of the residual (for the worst-case right-hand side vector) is completely determined by the eigenvalues of the matrix, and we have at least intuitive ideas of what constitutes good and bad eigenvalue distributions. The same reasoning shows that the A-norm of the error $e_k \equiv A^{-1}b - x_k$ in the CG algorithm satisfies

(1.6) $$\|e_k\|_A/\|A^{-1}b\|_A \leq \min_{p_k} \max_{i=1,\ldots,n} |p_k(\lambda_i)|,$$

and, again, this bound is sharp.

Thus, for Hermitian problems, we not only have good algorithms for finding the optimal approximation from the Krylov space (1.1), but we can also say just how good that approximation will be, based on simple properties of the coefficient matrix (i.e., the eigenvalues). It is therefore fair to say that the iterative solution of Hermitian linear systems is well understood—except for one thing. All of the above discussion assumes *exact arithmetic*. It is well known that in finite precision arithmetic the CG and MINRES algorithms do *not* find the optimal approximation from the Krylov space (1.1) or, necessarily, anything close to it! In fact, the CG algorithm originally lost favor partly because it did not behave the way exact arithmetic theory predicted [43]. More recent work [65, 71, 35, 34] has gone a long way toward explaining the behavior of the MINRES and CG algorithms in finite precision arithmetic, although open problems remain. This work is discussed in Chapter 4.

1.1.2. Non-Hermitian Matrices. While the methods of choice and convergence analysis for Hermitian problems are well understood in exact arithmetic and the results of Chapter 4 go a long way towards explaining the behavior of these methods in finite precision arithmetic, the state-of-the-art for *non-Hermitian* problems is not nearly so well developed. One difficulty is that no method is known for finding the optimal approximation from the space (1.1) while performing only $O(n)$ operations per iteration in addition to the matrix–vector multiplication. In fact, a theorem due to Faber and Manteuffel [45] shows that for most non-Hermitian matrices A there is no short recurrence for the optimal approximation from the Krylov space (1.1). To fully understand this result, one must consider the statement and hypotheses carefully, and we give more of these details in Chapter 6. Still, the current options for non-Hermitian problems are either to perform extra work ($O(nk)$ operations at step k) and use extra storage ($O(nk)$ words to perform k iterations) to find the optimal approximation from the Krylov space or to settle for a nonoptimal approximation. The (full) GMRES (generalized minimal residual) algorithm [119] (and other mathematically equivalent algorithms) finds the approximation of the form (1.1) for which the 2-norm of the residual is minimal at the cost of this extra work and storage, while other non-Hermitian iterative methods (e.g., BiCG [51], CGS [124], QMR [54], BiCGSTAB [134], restarted GMRES [119], hybrid GMRES [103], etc.) generate nonoptimal approximations.

Even if one does generate the optimal approximation of the form (1.1), we are still unable to answer question (i). That is, there is no known way to describe how good this optimal approximation will be (for the worst-case right-hand side vector) in terms of simple properties of the coefficient matrix. One might try an approach based on eigenvalues, as was done for the Hermitian case. Assume that A is diagonalizable and write an eigendecomposition of A as $A = V\Lambda V^{-1}$, where Λ is a diagonal matrix of eigenvalues and the columns of V are eigenvectors. Then it follows from (1.4) that the GMRES residual at step k satisfies

$$(1.7) \qquad \|r_k\| \leq \min_{p_k} \|Vp_k(\Lambda)V^{-1}b\| \leq \kappa(V) \min_{p_k} \max_{i=1,\ldots,n} |p_k(\lambda_i)| \, \|b\|,$$

where $\kappa(V) = \|V\| \cdot \|V^{-1}\|$ is the condition number of the eigenvector matrix. The scaling of the columns of V can be chosen to minimize this condition number. When A is a *normal* matrix, so that V can be taken to have condition number one, it turns out that the bound (1.7) is sharp, just as for the Hermitian case. Thus for normal matrices, the analysis of GMRES again reduces to a question of polynomial approximation—how well can one approximate zero on the set of (complex) eigenvalues using a kth-degree polynomial with value 1 at the origin? When A is nonnormal, however, the bound (1.7) may *not* be sharp. If the condition number of V is huge, the right-hand side of (1.7) may also be large, but this does not necessarily imply that GMRES converges poorly. It may simply mean that the bound (1.7) is a large overestimate of the actual GMRES residual norm. An interesting open question is to determine when an ill-conditioned eigenvector matrix implies poor convergence for GMRES and when it simply means that the bound (1.7) is a large overestimate. If eigenvalues are not the key, then one would like to be able to describe the behavior of GMRES in terms of some other characteristic properties of the matrix A. Some ideas along these lines are discussed in section 3.2.

Finally, since the full GMRES algorithm may be impractical if a fairly large number of iterations are required, one would like to have theorems relating the convergence of some nonoptimal methods (that do not require extra work and storage) to that of GMRES. Unfortunately, no fully satisfactory theorems of this kind are currently available, and this important open problem is discussed in Chapter 6.

1.1.3. Preconditioners. The tools used in the derivation of preconditioners are much more diverse than those applied to the study of iteration methods. There are some general results concerning comparison of preconditioners and optimality of preconditioners of certain forms (e.g., block-diagonal), and these are described in Chapter 10. Many of the most successful preconditioners, however, have been derived for special problem classes, where the origin of the problem suggests a particular type of preconditioner. Multigrid and domain decomposition methods are examples of this type of preconditioner and are discussed in Chapter 12. Still other preconditioners are designed for very specific physical problems, such as the transport equation. Since one cannot assume familiarity with every scientific application, a complete survey is impossible. Chapter 9 contains two example problems, but the problem of generalizing application-specific preconditioners to a broader setting remains an area of active research.

1.2. Notation.

We assume *complex* matrices and vectors throughout this book. The results for real problems are almost always the same, and we point out any differences that might be encountered. The symbol $\imath$ is used for $\sqrt{-1}$, and a superscript ^{H} denotes the Hermitian transpose ($A_{ij}^H = \bar{A}_{ji}$, where the overbar denotes the complex conjugate). The symbol $\|\cdot\|$ will always denote the 2-norm for vectors and the induced spectral norm for matrices. An arbitrary norm will be denoted $\|\|\cdot\|\|$.

The linear system (or sometimes the preconditioned linear system) under consideration is denoted $Ax = b$, where A is an n-by-n nonsingular matrix and b is a given n-vector. If x_k is an approximate solution then the residual $b - Ax_k$

is denoted as r_k and the error $A^{-1}b - x_k$ as e_k. The symbol ξ_j denotes the jth unit vector, i.e., the vector whose jth entry is 1 and whose other entries are 0, with the size of the vector being determined by the context.

A number of algorithms are considered, and these are first stated in the form most suitable for presentation. This does not always correspond to the best computational implementation. Algorithms enclosed in boxes are the recommended computational procedures, although details of the actual implementation (such as how to carry out a matrix–vector multiplication or how to solve a linear system with the preconditioner as coefficient matrix) are not included. Most of these algorithms are implemented in the Templates [10], with MATLAB, Fortran, and C versions. To see what is available in this package, type

```
mail netlib@ornl.gov
send index for templates
```

or explore the website: `http://www.netlib.org/templates`. Then, to obtain a specific version, such as the MATLAB version of the Templates, type

```
mail netlib@ornl.gov
send mltemplates.shar from templates
```

or download the appropriate file from the web. The reader is encouraged to experiment with the iterative methods described in this book, either through use of the Templates or another software package or through coding the algorithms directly.

1.3. Review of Relevant Linear Algebra.

1.3.1. Vector Norms and Inner Products.
We assume that the reader is already familiar with vector norms. Examples of vector norms are

- the Euclidean norm or 2-norm, $\|v\|_2 = \left(\sum_{i=1}^{n} |v_i|^2\right)^{1/2}$;

- the 1-norm, $\|v\|_1 = \sum_{i=1}^{n} |v_i|$; and

- the ∞-norm, $\|v\|_\infty = \max_{i=1,\ldots,n} |v_i|$.

From here on we will denote the 2-norm by, simply, $\| \cdot \|$. If $\|| \cdot \||$ is a vector norm and G is a nonsingular n-by-n matrix, then $\||v\||_{G^H G} \equiv \||Gv\||$ is also a vector norm. (This is sometimes denoted $\||v\||_G$ instead of $\||v\||_{G^H G}$, but we will use the latter notation and will refer to $\| \cdot \|_{G^H G}$ as the $G^H G$-norm in order to be consistent with standard notation used in describing the CG algorithm.)

The Euclidean norm is associated with an inner product:

$$\langle v, w \rangle \equiv w^H v \equiv \sum_{i=1}^{n} \bar{w}_i v_i.$$

We refer to this as the standard inner product. Similarly, the $G^H G$-norm is associated with an inner product:

$$\langle v, w \rangle_{G^H G} \equiv \langle v, G^H G w \rangle = \langle Gv, Gw \rangle.$$

By definition we have $\|v\|^2 = \langle v, v \rangle$, and it follows that $\|v\|^2_{G^H G} = \langle Gv, Gv \rangle = \langle v, v \rangle_{G^H G}$. If $\|\| \cdot \|\|$ is any norm associated with an inner product $\langle\langle \cdot, \cdot \rangle\rangle$, then there is a nonsingular matrix G such that

$$\|\|v\|\| = \|v\|_{G^H G} \quad \text{and} \quad \langle\langle v, w \rangle\rangle = \langle v, w \rangle_{G^H G}.$$

The i, j entry of $G^H G$ is $\langle\langle \xi_i, \xi_j \rangle\rangle$, where ξ_i and ξ_j are the unit vectors with one in position i and j, respectively, and zeros elsewhere.

1.3.2. Orthogonality. The vectors v and w are said to be *orthogonal* if $\langle v, w \rangle = 0$ and to be *orthonormal* if, in addition, $\|v\| = \|w\| = 1$. The vectors v and w are said to be $G^H G$-*orthogonal* if $\langle v, G^H G w \rangle = 0$.

The $G^H G$-*projection* of a vector v in the direction w is

$$\frac{\langle v, G^H G w \rangle}{\langle w, G^H G w \rangle} w,$$

and to minimize the $G^H G$-norm of v in the direction w, one subtracts off the $G^H G$-projection of v onto w. That is, if

$$\hat{v} = v - \frac{\langle v, G^H G w \rangle}{\langle w, G^H G w \rangle} w,$$

then of all vectors of the form $v - \alpha w$ where α is a scalar, $\hat{v}$ has the smallest $G^H G$-norm.

An n-by-n complex matrix with orthonormal columns is called a *unitary* matrix. For a unitary matrix Q, we have $Q^H Q = Q Q^H = I$, where I is the n-by-n identity matrix. If the matrix Q is real, then it can also be called an *orthogonal* matrix.

Given a set of linearly independent vectors $\{v_1, \ldots, v_n\}$, one can construct an orthonormal set $\{u_1, \ldots, u_n\}$ using the *Gram–Schmidt* procedure:

$$u_1 = v_1 / \|v_1\|,$$

$$\tilde{u}_k = v_k - \sum_{i=1}^{k-1} \langle v_k, u_i \rangle u_i, \quad u_k = \tilde{u}_k / \|\tilde{u}_k\|, \quad k = 1, \ldots, n.$$

In actual computations, a mathematically equivalent procedure called the *modified Gram–Schmidt* method is often used:

Modified Gram–Schmidt Algorithm.

Set $u_1 = v_1 / \|v_1\|$. For $k = 1, \ldots, n$,

$\quad \tilde{u}_k = v_k$. For $i = 1, \ldots, k-1$, $\quad \tilde{u}_k \longleftarrow \tilde{u}_k - \langle \tilde{u}_k, u_i \rangle u_i$.

$\quad u_k = \tilde{u}_k / \|\tilde{u}_k\|$.

Here, instead of computing the projection of v_k onto each of the basis vectors u_i, $i = 1, \ldots, k - 1$, the next basis vector is formed by first subtracting off the projection of v_k in the direction of one of the basis vectors and then subtracting off the projection of the *new* vector $\tilde{u}_k$ in the direction of another basis vector, etc. The modified Gram–Schmidt procedure forms the core of many iterative methods for solving linear systems.

If U_k is the matrix whose columns are the orthonormal vectors $u_1, \ldots, u_k$, then the closest vector to a given vector v from the space span$\{u_1, \ldots, u_k\}$ is the *projection* of v onto this space,

$$U_k U_k^H v.$$

1.3.3. Matrix Norms.   Let M_n denote the set of n-by-n complex matrices.

DEFINITION 1.3.1. *A function* $||| \cdot ||| : M_n \to \mathbf{R}$ *is called a* matrix norm *if, for all* $A, B \in M_n$ *and all complex scalars* c,

1. $|||A||| \geq 0$ *and* $|||A||| = 0$ *if and only if* $A = 0$;

2. $|||cA||| = |c| \cdot |||A|||$;

3. $|||A + B||| \leq |||A||| + |||B|||$;

4. $|||AB||| \leq |||A||| \cdot |||B|||$.

Example. The *Frobenius* norm defined by

$$\|A\|_F^2 = \sum_{i,j=1}^{n} |a_{i,j}|^2$$

is a matrix norm because, in addition to properties 1–3, we have

$$\|AB\|_F^2 = \sum_{i,j=1}^{n} \left| \sum_{k=1}^{n} a_{ik} b_{kj} \right|^2 \leq \sum_{i,j=1}^{n} \left(\sum_{k=1}^{n} |a_{ik}|^2 \right) \left(\sum_{\ell=1}^{n} |b_{\ell,j}|^2 \right)$$

$$= \left(\sum_{i,k=1}^{n} |a_{ik}|^2 \right) \left(\sum_{\ell,j=1}^{n} |b_{\ell,j}|^2 \right) = \|A\|_F^2 \cdot \|B\|_F^2.$$

This is just the Cauchy–Schwarz inequality.

DEFINITION 1.3.2. *Let* $||| \cdot |||$ *be a vector norm on* $\mathbf{C}^n$. *The* induced norm, *also denoted* $||| \cdot |||$, *is defined on* M_n *by*

$$|||A||| = \max_{|||y|||=1} |||Ay|||.$$

The "max" in the above definition could be replaced by "sup." The two are equivalent since $|||Ay|||$ is a continuous function of y and the unit ball in $\mathbf{C}^n$, being a compact set, contains the vector for which the sup is attained. Another equivalent definition is

$$|||A||| = \max_{y \neq 0} |||Ay|||/|||y|||.$$

The norm $\|\cdot\|_1$ induced on M_n by the 1-norm for vectors is

$$\|A\|_1 = \max_j \sum_{i=1}^n |a_{ij}|,$$

the maximal absolute column sum of A. To see this, write A in terms of its columns $A = [\mathbf{a_1}, \ldots, \mathbf{a_n}]$. Then

$$
\begin{aligned}
\|Ay\|_1 &= \left\| \sum_{i=1}^n \mathbf{a_i} y_i \right\|_1 \leq \sum_{i=1}^n \|\mathbf{a_i} y_i\|_1 = \sum_{i=1}^n \|\mathbf{a_i}\|_1 \cdot |y_i| \\
&\leq \max_i \|\mathbf{a_i}\|_1 \cdot \sum_{i=1}^n |y_i| = \|\mathbf{A}\|_1 \cdot \|\mathbf{y}\|_1.
\end{aligned}
$$

Thus, $\max_{\|y\|_1=1} \|Ay\|_1 \leq \|A\|_1$. But if y is the unit vector with a 1 in the position of the column of A having the greatest 1-norm and zeros elsewhere, then $\|Ay\|_1 = \|A\|_1$, so we also have $\max_{\|y\|_1=1} \|Ay\|_1 \geq \|A\|_1$.

The norm $\|\cdot\|_\infty$ induced on M_n by the ∞-norm for vectors is

$$\|A\|_\infty = \max_i \sum_{j=1}^n |a_{ij}|,$$

the largest absolute row sum of A. To see this, first note that

$$
\begin{aligned}
\|Ay\|_\infty &= \max_i \left| \sum_{j=1}^n a_{ij} y_j \right| \leq \max_i \sum_{j=1}^n |a_{ij} y_j| \leq \max_j |y_j| \cdot \max_i \sum_{j=1}^n |a_{ij}| \\
&\leq \|y\|_\infty \cdot \|A\|_\infty,
\end{aligned}
$$

and hence $\max_{\|y\|_\infty=1} \|Ay\|_\infty \leq \|A\|_\infty$. On the other hand, suppose the kth row of A has the largest absolute row sum. Take y to be a vector of ± 1's with the sign of each entry j matching that of a_{kj}. Then the kth entry of Ay is the sum of the absolute values of the entries in row k of A, so we have $\|Ay\|_\infty \geq \|A\|_\infty$.

The norm $\|\cdot\|_2$ induced on M_n by the 2-norm for vectors is

$$\|A\|_2 = \max\{\sqrt{\lambda} : \lambda \text{ is an eigenvalue of } A^H A\}.$$

To see this, recall the variational characterization of eigenvalues of a Hermitian matrix:

$$\lambda_{max}(A^H A) = \max_{y \neq 0} \frac{y^H A^H A y}{y^H y} = \max_{\|y\|=1} \|Ay\|^2.$$

This matrix norm will also be denoted, simply, as $\| \cdot \|$ from here on.

THEOREM 1.3.1. *If* $\vert\vert\vert \cdot \vert\vert\vert$ *is a matrix norm on* M_n *and if* $G \in M_n$ *is nonsingular, then*

$$\vert\vert\vert A \vert\vert\vert_{G^H G} \equiv \vert\vert\vert GAG^{-1} \vert\vert\vert$$

is a matrix norm. If $\vert\vert\vert \cdot \vert\vert\vert$ *is induced by a vector norm* $\vert\vert\vert \cdot \vert\vert\vert$, *then* $\vert\vert\vert \cdot \vert\vert\vert_{G^H G}$ *is induced by the vector norm* $\vert\vert\vert \cdot \vert\vert\vert_{G^H G}$.

Proof. Axioms 1–3 of Definition 1.3.1 are easy to verify, and axiom 4 follows from

$$\vert\vert\vert AB \vert\vert\vert_{G^H G} = \vert\vert\vert GABG^{-1} \vert\vert\vert = \vert\vert\vert (GAG^{-1})(GBG^{-1}) \vert\vert\vert$$

$$\leq \vert\vert\vert GAG^{-1} \vert\vert\vert \cdot \vert\vert\vert GBG^{-1} \vert\vert\vert = \vert\vert\vert A \vert\vert\vert_{G^H G} \cdot \vert\vert\vert B \vert\vert\vert_{G^H G}.$$

If $\vert\vert\vert A \vert\vert\vert = \max_{y \neq 0} \vert\vert\vert Ay \vert\vert\vert / \vert\vert\vert y \vert\vert\vert$, then

$$
\begin{aligned}
\vert\vert\vert A \vert\vert\vert_{G^H G} &= \max_{y \neq 0} \vert\vert\vert GAG^{-1} y \vert\vert\vert / \vert\vert\vert y \vert\vert\vert \\
&= \max_{w \neq 0} \vert\vert\vert GAw \vert\vert\vert / \vert\vert\vert Gw \vert\vert\vert \\
&= \max_{w \neq 0} \vert\vert\vert Aw \vert\vert\vert_{G^H G} / \vert\vert\vert w \vert\vert\vert_{G^H G},
\end{aligned}
$$

so $\vert\vert\vert \cdot \vert\vert\vert_{G^H G}$ is the matrix norm induced by the vector norm $\vert\vert\vert \cdot \vert\vert\vert_{G^H G}$. □

1.3.4. The Spectral Radius. The *spectral radius*, or largest absolute value of an eigenvalue of a matrix, is of importance in the analysis of certain iterative methods.

DEFINITION 1.3.3. *The spectral radius* $\rho(A)$ *of a matrix* $A \in M_n$ *is*

$$\rho(A) = \max\{|\lambda| : \lambda \text{ is an eigenvalue of } A\}.$$

THEOREM 1.3.2. *If* $\vert\vert\vert \cdot \vert\vert\vert$ *is any matrix norm and* $A \in M_n$, *then* $\rho(A) \leq \vert\vert\vert A \vert\vert\vert$.

Proof. Let λ be an eigenvalue of A with $|\lambda| = \rho(A)$, and let v be the corresponding eigenvector. Let V be the matrix in M_n each of whose columns is v. Then $AV = \lambda V$, and if $\vert\vert\vert \cdot \vert\vert\vert$ is any matrix norm,

$$|\lambda| \cdot \vert\vert\vert V \vert\vert\vert = \vert\vert\vert \lambda V \vert\vert\vert = \vert\vert\vert AV \vert\vert\vert \leq \vert\vert\vert A \vert\vert\vert \cdot \vert\vert\vert V \vert\vert\vert.$$

Since $\vert\vert\vert V \vert\vert\vert > 0$, it follows that $\rho(A) \leq \vert\vert\vert A \vert\vert\vert$. □

THEOREM 1.3.3. *Let* $A \in M_n$ *and* $\epsilon > 0$ *be given. There is a matrix norm* $\vert\vert\vert \cdot \vert\vert\vert$ *induced by a certain vector norm such that*

$$\rho(A) \leq \vert\vert\vert A \vert\vert\vert \leq \rho(A) + \epsilon.$$

Proof. The proof of this theorem will use the Schur triangularization, which is stated as a theorem in the next section. According to this theorem, there is a unitary matrix Q and an upper triangular matrix U whose diagonal entries are the eigenvalues of A such that $A = QUQ^H$. Set $D_t = \mathrm{diag}(t, t^2, \ldots, t^n)$ and note that

$$D_t U D_t^{-1} = \begin{pmatrix} \lambda_1 & t^{-1}u_{12} & t^{-2}u_{13} & \cdots & t^{-n+1}u_{1n} \\ & \lambda_2 & t^{-1}u_{23} & \cdots & t^{-n+2}u_{2n} \\ & & \lambda_3 & \cdots & t^{-n+3}u_{3n} \\ & & & \ddots & \vdots \\ & & & & \lambda_n \end{pmatrix}.$$

For t sufficiently large, the sum of absolute values of all off-diagonal elements in a column is less than ϵ, so $\|D_t U D_t^{-1}\|_1 \leq \rho(A) + \epsilon$. Thus if we define the matrix norm $||| \cdot |||$ by

$$|||B||| \equiv \|D_t Q^H B Q D_t^{-1}\|_1$$

for any matrix $B \in M_n$, then we will have $|||A||| = \|D_t U D_t^{-1}\|_1 \leq \rho(A) + \epsilon$. It follows from Theorem 1.3.1 that $||| \cdot |||$ is a matrix norm since

$$|||B||| = \|GBG^{-1}\|_1, \quad G = D_t Q^H,$$

and that it is induced by the vector norm

$$|||y||| \equiv \|D_t Q^H y\|_1. \quad \square$$

To study the convergence of simple iteration, we will be interested in conditions under which the powers of a matrix A converge to the zero matrix.

THEOREM 1.3.4. *Let $A \in M_n$. Then $\lim_{k \to \infty} A^k = 0$ if and only if $\rho(A) < 1$.*

Proof. First suppose $\lim_{k \to \infty} A^k = 0$. Let λ be an eigenvalue of A with eigenvector v. Since $A^k v = \lambda^k v \to 0$ as $k \to \infty$, this implies $|\lambda| < 1$. Conversely, if $\rho(A) < 1$, then by Theorem 1.3.3 there is a matrix norm $||| \cdot |||$ such that $|||A||| < 1$. It follows that $|||A^k||| \leq |||A|||^k \to 0$ as $k \to \infty$. Since all matrix norms on M_n are equivalent, this implies, for instance, that $\|A^k\|_F \to 0$ as $k \to \infty$, which implies that all entries of A^k must approach 0. $\quad \square$

COROLLARY 1.3.1. *Let $||| \cdot |||$ be a matrix norm on M_n. Then*

$$\rho(A) = \lim_{k \to \infty} |||A^k|||^{1/k}$$

for all $A \in M_n$.

Proof. Since $\rho(A)^k = \rho(A^k) \leq |||A^k|||$, we have $\rho(A) \leq |||A^k|||^{1/k}$, for all $k = 1, 2, \ldots$. For any $\epsilon > 0$, the matrix $\tilde{A} \equiv [\rho(A) + \epsilon]^{-1} A$ has spectral radius strictly less than one and so $|||\tilde{A}^k||| \to 0$ as $k \to \infty$. There is some number $K = K(\epsilon)$ such that $|||\tilde{A}^k||| < 1$ for all $k \geq K$, and this is just the statement that $|||A^k||| \leq [\rho(A) + \epsilon]^k$ or $|||A^k|||^{1/k} \leq \rho(A) + \epsilon$ for all $k \geq K$. We thus have $\rho(A) \leq |||A^k|||^{1/k} \leq \rho(A) + \epsilon$, for all $k \geq K(\epsilon)$, and since this holds for any $\epsilon > 0$, it follows that $\lim_{k \to \infty} |||A^k|||^{1/k}$ exists and is equal to $\rho(A)$. $\quad \square$

1.3.5. Canonical Forms and Decompositions. Matrices can be reduced to a number of different forms through similarity transformations, and they can be factored in a variety of ways. These forms are often useful in the analysis of numerical algorithms. Several such canonical representations and decompositions are described here, without proofs of their existence or algorithms for their computation.

THEOREM 1.3.5 (Jordan form). *Let A be an n-by-n matrix. There is a nonsingular matrix S such that*

$$
(1.8) \qquad A = S \begin{bmatrix} J_{n_1}(\lambda_1) & & & \\ & J_{n_2}(\lambda_2) & & \\ & & \ddots & \\ & & & J_{n_m}(\lambda_m) \end{bmatrix} S^{-1} = SJS^{-1},
$$

where

$$
(1.9) \qquad J_{n_i}(\lambda_i) = \begin{pmatrix} \lambda_i & 1 & & \\ & \lambda_i & \ddots & \\ & & \ddots & 1 \\ & & & \lambda_i \end{pmatrix}, \qquad n_i \times n_i,
$$

and $\sum_{i=1}^{m} n_i = n$.

The matrix J is called the *Jordan form* of A. The columns of S are called *principal vectors*. The number m of Jordan blocks is the number of independent eigenvectors of A.

The matrix A is *diagonalizable* if and only if $m = n$, and in this case the columns of S are called eigenvectors. The set of diagonalizable matrices is *dense* in the set of all matrices. To see this, consider perturbing the diagonal entries of a Jordan block by arbitrarily tiny amounts so that they are all different. The matrix then has distinct eigenvalues, and any matrix with distinct eigenvalues is diagonalizable.

In the special case when A is diagonalizable and the columns of S are orthogonal, the matrix A is said to be a *normal* matrix.

DEFINITION 1.3.4. *A matrix A is* normal *if it can be written in the form $A = Q\Lambda Q^H$, where Λ is a diagonal matrix and Q is a unitary matrix.*

A matrix is normal if and only if it commutes with its Hermitian transpose: $AA^H = A^H A$. Any Hermitian matrix is normal.

It can be shown by induction that the kth power of a j-by-j Jordan block corresponding to the eigenvalue λ is given by

$$
J^k =
\begin{pmatrix}
\lambda^k & \binom{k}{1}\lambda^{k-1} & \binom{k}{2}\lambda^{k-2} & \cdots & \binom{k}{j-1}\lambda^{k-j+1} \\
 & \lambda^k & \binom{k}{1}\lambda^{k-1} & \cdots & \binom{k}{j-2}\lambda^{k-j+2} \\
 & & \lambda^k & \cdots & \binom{k}{j-3}\lambda^{k-j+3} \\
 & & & \ddots & \vdots \\
 & & & & \lambda^k
\end{pmatrix},
$$

where $\binom{k}{i}$ is taken to be 0 if $i > k$. The 2-norm of J^k satisfies

$$
\|J^k\| \sim \binom{k}{j-1}[\rho(J)]^{k-j+1}, \quad k \to \infty,
$$

where the symbol $\sim$ means that, asymptotically, the left-hand side behaves like the right-hand side. The 2-norm of an arbitrary matrix A^k satisfies

$$
\|A^k\| \sim \nu \binom{k}{j-1}[\rho(A)]^{k-j+1},
$$

where j is the largest order of all diagonal submatrices J_r of the Jordan form with $\rho(J_r) = \rho(A)$ and ν is a positive constant.

THEOREM 1.3.6 (Schur form). *Let A be an n-by-n matrix with eigenvalues $\lambda_1, \ldots, \lambda_n$ in any prescribed order. There is a unitary matrix Q such that $A = QUQ^H$, where U is an upper triangular matrix and $U_{i,i} = \lambda_i$.*

Note that while the transformation S taking a matrix to its Jordan form may be extremely ill conditioned (that is, S in Theorem 1.3.5 may be nearly singular), the transformation to upper triangular form is perfectly conditioned (Q in Theorem 1.3.6 is unitary). Consequently, the Schur form often proves more useful in numerical analysis.

The Schur form is not unique, since the diagonal entries of U may appear in any order, and the entries of the upper triangle may be very different depending on the ordering of the diagonal entries. For example, the upper triangular matrices

$$
U_1 = \begin{pmatrix} 1 & 1 & 1 \\ 0 & 2 & 1 \\ 0 & 0 & 3 \end{pmatrix} \quad \text{and} \quad U_2 = \begin{pmatrix} 2 & -1 & \sqrt{2} \\ 0 & 1 & 0 \\ 0 & 0 & 3 \end{pmatrix}
$$

are two Schur forms of the same matrix, since they are unitarily equivalent via

$$
Q = \frac{1}{\sqrt{2}} \begin{pmatrix} 1 & 1 & 0 \\ 1 & -1 & 0 \\ 0 & 0 & \sqrt{2} \end{pmatrix}.
$$

THEOREM 1.3.7 (*LU decomposition*). *Let A in M_n be nonsingular. Then A can be factored in the form*

$$(1.10) \qquad\qquad\qquad A = PLU,$$

where P is a permutation matrix, L is lower triangular, and U is upper triangular.

The *LU* decomposition is a standard direct method for solving a linear system $Ax = b$. Factor the matrix A into the form PLU, solve $Ly = P^H b$ (since $P^H = P^{-1}$), and then solve $Ux = y$. Unfortunately, even if the matrix A is sparse, the factors L and U are usually dense (at least within a band about the diagonal). In general, the work required to compute the *LU* decomposition is $O(n^3)$ and the work to backsolve with the computed *LU* factors is $O(n^2)$. It is for this reason that iterative linear system solvers, which may require far less work and storage, are important. In Chapter 11, we discuss the use of "incomplete" *LU* decompositions as preconditioners for iterative methods. The idea is to drop entries of L and U that are small or are outside of certain positions.

For certain matrices A, the permutation matrix P in (1.10) is not necessary; that is, P can be taken to be the identity. For Hermitian positive definite matrices, for instance, no permutation is required for the *LU* decomposition, and if L and U^H are taken to have the same diagonal entries, then this decomposition becomes $A = LL^H$. This is sometimes referred to as the *Cholesky decomposition*.

Another direct method for solving linear systems or least squares problems is the QR decomposition.

THEOREM 1.3.8 (QR decomposition). *Let A be an m-by-n matrix with $m \geq n$. There is an m-by-n matrix Q with orthonormal columns and an n-by-n upper triangular matrix R such that $A = QR$. Columns can be added to the matrix Q to form an m-by-m unitary matrix $\hat{Q}$ such that $A = \hat{Q}\hat{R}$, where $\hat{R}$ is an m-by-n matrix with R as its top n-by-n block and zeros elsewhere.*

One way to compute the QR decomposition of a matrix A is to apply the modified Gram–Schmidt algorithm of section 1.3.2 to the columns of A. Another way is to apply a sequence of unitary matrices to A to transform it to an upper triangular matrix. Since the product of unitary matrices is unitary and the inverse of a unitary matrix is unitary, this also gives a QR factorization of A. If A is square and nonsingular and the diagonal elements of R are taken to be positive, then the Q and R factors are unique, so this gives the same QR factorization as the modified Gram–Schmidt algorithm. Unitary matrices that are often used in the QR decomposition are *reflections* (Householder transformations) and, for matrices with special structures, *rotations* (Givens transformations). A number of iterative linear system solvers apply Givens rotations to a smaller upper Hessenberg matrix in order to solve a least squares problem with this smaller coefficient matrix. Once an m-by-n matrix has been factored in the form $\hat{Q}\hat{R}$, a least squares problem—find y to minimize

$\|\hat{Q}\hat{R}y - b\|$—can be solved by solving the upper triangular system $Ry = Q^Hb$.

THEOREM 1.3.9 (singular value decomposition). *If A is an m-by-n matrix with rank k, then it can be written in the form*

(1.11) $$A = V\Sigma W^H,$$

where V is an m-by-m unitary matrix, W is an n-by-n unitary matrix, and Σ is an m-by-n matrix with $\sigma_{i,j} = 0$ for all $i \neq j$ and $\sigma_{11} \geq \sigma_{22} \geq \cdots \geq \sigma_{kk} > \sigma_{k+1,k+1} = \cdots = \sigma_{qq} = 0$, where $q = \min\{m,n\}$.

The numbers $\sigma_{ii} \equiv \sigma_i$, known as *singular values* of A, are the nonnegative square roots of the eigenvalues of AA^H. The columns of V, known as *left singular vectors* of A, are eigenvectors of AA^H; the columns of W, known as *right singular vectors* of A, are eigenvectors of $A^H A$.

Using the singular value decomposition and the Schur form, we are able to define a certain measure of the *departure from normality* of a matrix. It is the difference between the sum of squares of the singular values and the sum of squares of the eigenvalues, and it is also equal to the sum of squares of the entries in the strict upper triangle of a Schur form. For a normal matrix, each of these quantities is, of course, zero.

THEOREM 1.3.10. *Let $A \in M_n$ have eigenvalues $\lambda_1, \ldots, \lambda_n$ and singular values $\sigma_1, \ldots, \sigma_n$, and let $A = QUQ^H$ be a Schur decomposition of A. Let Λ denote the diagonal of U, consisting of the eigenvalues of A, in some order, and let T denote the strict upper triangle of U. Then*

$$\|A\|_F^2 = \sum_{i=1}^{n} \sigma_i^2 = \|\Lambda\|_F^2 + \|T\|_F^2.$$

Proof. From the definition of the Frobenius norm, it is seen that $\|A\|_F^2 = \text{tr}(A^H A)$, where $\text{tr}(\cdot)$ denotes the trace, i.e., the sum of the diagonal entries. If $A = V\Sigma W^H$ is the singular value decomposition of A, then

$$\text{tr}(A^H A) = \text{tr}(W\Sigma^2 W^H) = \text{tr}(\Sigma^2) = \sum_{i=1}^{n} \sigma_i^2.$$

If $A = QUQ^H$ is a Schur form of A, then it is also clear that

$$\text{tr}(A^H A) = \text{tr}(Q^H U^H UQ) = \text{tr}(U^H U) = \|U\|_F^2 = \|\Lambda\|_F^2 + \|T\|_F^2. \quad \square$$

DEFINITION 1.3.5. *The Frobenius norm of the strict upper triangle of a Schur form of A is called the* departure from normality *of A with respect to the Frobenius norm.*

1.3.6. Eigenvalues and the Field of Values.   We will see later that the *eigenvalues* of a normal matrix provide all of the essential information about that matrix, as far as iterative linear system solvers are concerned. There is no corresponding simple set of characteristics of a nonnormal matrix that provide such complete information, but the *field of values* captures certain important properties.

We begin with the useful theorem of Gerschgorin for locating the eigenvalues of a matrix.

THEOREM 1.3.11 (Gerschgorin). *Let A be an n-by-n matrix and let*

$$(1.12) \qquad R_i(A) = \sum_{\substack{j=1 \\ j\neq i}}^{n} |a_{i,j}|, \quad i = 1, \ldots, n$$

denote the sum of the absolute values of all off-diagonal entries in row i. Then all eigenvalues of A are located in the union of disks

$$(1.13) \qquad \bigcup_{i=1}^{n} \{z \in \mathbf{C} : |z - a_{i,i}| \leq R_i(A)\}.$$

Proof. Let λ be an eigenvalue of A with corresponding eigenvector v. Let v_p be the element of v with largest absolute value, $|v_p| \geq \max_{i \neq p} |v_i|$. Then since $Av = \lambda v$, we have

$$(Av)_p = \lambda v_p = \sum_{j=1}^{n} a_{p,j} v_j$$

or, equivalently,

$$v_p(\lambda - a_{p,p}) = \sum_{\substack{j=1 \\ j\neq p}}^{n} a_{p,j} v_j.$$

From the triangle inequality it follows that

$$|v_p||\lambda - a_{p,p}| = \left| \sum_{\substack{j=1 \\ j\neq p}}^{n} a_{p,j} v_j \right| \leq \sum_{\substack{j=1 \\ j\neq p}}^{n} |a_{p,j}| \, |v_j| \leq |v_p| R_p(A).$$

Since $|v_p| > 0$, it follows that $|\lambda - a_{p,p}| \leq R_p(A)$; that is, the eigenvalue λ lies in the Gerschgorin disk for the row corresponding to its eigenvector's largest entry, and hence all of the eigenvalues lie in the union of the Gerschgorin disks. $\square$

It can be shown further that if a union of k of the Gerschgorin disks forms a connected region that is disjoint from the remaining $n - k$ disks, then that region contains exactly k of the eigenvalues of A.

Since the eigenvalues of A are the same as those of A^H, the following corollary is immediate.

COROLLARY 1.3.2. *Let A be an n-by-n matrix and let*

$$(1.14) \qquad C_j(A) = \sum_{\substack{i=1 \\ i\neq j}}^{n} |a_{i,j}|, \quad j = 1, \ldots, n$$

*denote the sum of the absolute values of all off-diagonal entries in column j.
Then all eigenvalues of A are located in the union of disks*

(1.15)
$$\bigcup_{j=1}^{n} \{z \in \mathbf{C} : |z - a_{j,j}| \leq C_j(A)\}.$$

A matrix is said to be (rowwise) *diagonally dominant* if the absolute value
of each diagonal entry is strictly greater than the sum of the absolute values
of the off-diagonal entries in its row. In this case, the matrix is nonsingular,
since the Gerschgorin disks in (1.13) do not contain the origin. If the absolute
value of each diagonal entry is greater than or equal to the sum of the absolute
values of the off-diagonal entries in its row, then the matrix is said to be
weakly (rowwise) *diagonally dominant*. Analogous definitions of (columnwise)
diagonal dominance and weak diagonal dominance can be given.

We say that a Hermitian matrix is *positive definite* if its eigenvalues are all
positive. (Recall that the eigenvalues of a Hermitian matrix are all real.) A
diagonally dominant Hermitian matrix with positive diagonal entries is positive
definite. A Hermitian matrix is *positive semidefinite* if its eigenvalues are all
nonnegative. A weakly diagonally dominant Hermitian matrix with positive
diagonal entries is positive semidefinite.

Another useful theorem for obtaining bounds on the eigenvalues of a
Hermitian matrix is the Cauchy interlace theorem, which we state here without
proof. For a proof see, for instance, [81] or [115].

THEOREM 1.3.12 (Cauchy interlace theorem). *Let A be an n-by-n Hermi-
tian matrix with eigenvalues $\lambda_1 \leq \cdots \leq \lambda_n$, and let H be any m-by-m principal
submatrix of A (obtained by deleting $n-m$ rows and the corresponding columns
from A), with eigenvalues $\mu_1 \leq \cdots \leq \mu_m$. Then for each $i = 1, \ldots, m$ we have*

$$\lambda_i \leq \mu_i \leq \lambda_{i+n-m}.$$

For non-Hermitian matrices, the *field of values* is sometimes a more useful
concept than the eigenvalues.

DEFINITION 1.3.6. *The* field of values *of $A \in M_n$ is*

$$\mathcal{F}(A) = \{y^H A y : y \in \mathbf{C}^n, y^H y = 1\}.$$

This set is also called the *numerical range*. An equivalent definition is

$$\mathcal{F}(A) = \left\{ \frac{y^H A y}{y^H y} : y \in \mathbf{C}^n, y \neq 0 \right\}.$$

The field of values is a *compact* set in the complex plane, since it is the
continuous image of a compact set—the surface of the Euclidean ball. It can
also be shown to be a *convex* set. This is known as the *Toeplitz–Hausdorff*
theorem. See, for example, [81] for a proof. The *numerical radius* $\nu(A)$ is the
largest absolute value of an element of $\mathcal{F}(A)$:

$$\nu(A) \equiv \max\{|z| : z \in \mathcal{F}(A)\}.$$

If A is an n-by-n matrix and α is a complex scalar, then

(1.16)
$$\mathcal{F}(A + \alpha I) = \mathcal{F}(A) + \alpha,$$

since

$$
\begin{aligned}
\mathcal{F}(A + \alpha I) &= \{y^H(A + \alpha I)y : \ y^H y = 1\} \\
&= \{y^H Ay + \alpha y^H y : \ y^H y = 1\} \\
&= \{y^H Ay : \ y^H y = 1\} + \alpha = \mathcal{F}(A) + \alpha.
\end{aligned}
$$

Also,

(1.17)
$$\mathcal{F}(\alpha A) = \alpha \mathcal{F}(A),$$

since

$$
\begin{aligned}
\mathcal{F}(\alpha A) &= \{y^H \alpha Ay : \ y^H y = 1\} \\
&= \{\alpha y^H Ay : \ y^H y = 1\} = \alpha \mathcal{F}(A).
\end{aligned}
$$

For any n-by-n matrix A, $\mathcal{F}(A)$ contains the eigenvalues of A, since $v^H Av = \lambda v^H v = \lambda$ if λ is an eigenvalue and v is a corresponding normalized eigenvector. Also, if Q is any unitary matrix, then $\mathcal{F}(Q^H AQ) = \mathcal{F}(A)$, since every value $y^H Q^H AQy$ with $y^H y = 1$ in $\mathcal{F}(Q^H AQ)$ corresponds to a value $w^H Aw$ with $w = Qy$, $w^H w = 1$ in $\mathcal{F}(A)$, and vice versa.

For *normal* matrices the field of values is the *convex hull* of the spectrum. To see this, write the eigendecomposition of a normal matrix A in the form $A = Q\Lambda Q^H$, where Q is unitary and $\Lambda = \text{diag}\{\lambda_1, \ldots, \lambda_n\}$. By the unitary similarity invariance property, $\mathcal{F}(A) = \mathcal{F}(\Lambda)$. Since

$$y^H \Lambda y = \sum_{i=1}^{n} \bar{y}_i y_i \lambda_i = \sum_{i=1}^{n} |y_i|^2 \lambda_i,$$

it follows that $\mathcal{F}(\Lambda)$ is just the set of all convex combinations of the eigenvalues $\lambda_1, \ldots, \lambda_n$.

For a general matrix A, let $H(A) = \frac{1}{2}(A + A^H)$ denote the Hermitian part of A. Then

(1.18)
$$\mathcal{F}(H(A)) = \text{Re}(F(A)).$$

To see this, note that for any vector $y \in \mathbf{C}^n$,

$$y^H H(A)y = \frac{1}{2}(y^H Ay + y^H A^H y) = \frac{1}{2}(y^H Ay + \overline{y^H Ay}) = \text{Re}(y^H Ay).$$

Thus each point in $\mathcal{F}(H(A))$ is of the form $\text{Re}(z)$ for some $z \in \mathcal{F}(A)$ and vice versa.

The analogue of Gerschgorin's theorem for the field of values is as follows.

THEOREM 1.3.13. *Let A be an n-by-n matrix and let*

$$R_i(A) = \sum_{\substack{j=1 \\ j \neq i}}^{n} |a_{i,j}|, \quad i = 1, \ldots, n,$$

$$C_j(A) = \sum_{\substack{i=1 \\ i \neq j}}^{n} |a_{i,j}|, \quad j = 1, \ldots, n.$$

Then the field of values of A is contained in

$$(1.19) \qquad Co\left(\bigcup_{i=1}^{n} \left\{ z \in \mathbf{C} : \ |z - a_{i,i}| \leq \frac{1}{2}(R_i(A) + C_i(A)) \right\} \right),$$

where $Co(\cdot)$ denotes the convex hull.

Proof. First note that since the real part of $\mathcal{F}(A)$ is equal to $\mathcal{F}(H(A))$ and since $\mathcal{F}(H(A))$ is the convex hull of the eigenvalues of $H(A)$, it follows from Gerschgorin's theorem applied to $H(A)$ that

$$(1.20)$$

$$\mathrm{Re}(\mathcal{F}(A)) \subset Co\left(\bigcup_{i=1}^{n} \left\{ z \in \mathbf{R} : \ |z - \mathrm{Re}(a_{i,i})| \leq \frac{1}{2}(R_i(A) + C_i(A)) \right\} \right).$$

Let $G_F(A)$ denote the set in (1.19). If $G_F(A)$ is contained in the open right half-plane $\{z : \mathrm{Re}(z) > 0\}$, then $\mathrm{Re}(a_{i,i}) > \frac{1}{2}(R_i(A) + C_i(A))$ for all i, and hence the set on the right in (1.20) is contained in the open right half-plane. Since $\mathcal{F}(A)$ is convex, it follows that $\mathcal{F}(A)$ lies in the open right half-plane.

Now suppose only that $G_F(A)$ is contained in some open half-plane about the origin. Since $G_F(A)$ is convex, this is equivalent to the condition $0 \notin G_F(A)$. Then there is some $\theta \in [0, 2\pi)$ such that $e^{i\theta}G_F(A) = G_F(e^{i\theta}A)$ is contained in the open right half-plane. It follows from the previous argument that $\mathcal{F}(e^{i\theta}A) = e^{i\theta}\mathcal{F}(A)$ lies in the open right half-plane, and hence $0 \notin \mathcal{F}(A)$.

Finally, for any complex number α, if $\alpha \notin G_F(A)$ then $0 \notin G_F(A - \alpha I)$, and the previous argument implies that $0 \notin \mathcal{F}(A - \alpha I)$. Using (1.16), it follows that $\alpha \notin \mathcal{F}(A)$. Therefore, $\mathcal{F}(A) \subset G_F(A)$. $\square$

The following procedure can be used to approximate the field of values numerically. First note that since $\mathcal{F}(A)$ is convex and compact it is necessary only to compute the boundary. If many well-spaced points are computed around the boundary, then the convex hull of these points is a polygon $p(A)$ that is contained in $\mathcal{F}(A)$, while the intersection of the half-planes determined by the support lines at these points is a polygon $P(A)$ that contains $\mathcal{F}(A)$.

To compute points around the boundary of $\mathcal{F}(A)$, first note from (1.18) that the rightmost point in $\mathcal{F}(A)$ has real part equal to the rightmost point in $\mathcal{F}(H(A))$, which is the largest eigenvalue of $H(A)$. If we compute the largest eigenvalue λ_{max} of $H(A)$ and the corresponding unit eigenvector v, then $v^H A v$

is a boundary point of $\mathcal{F}(A)$ and the vertical line $\{\lambda_{max} + t\imath, \ t \in \mathbf{R}\}$, is a support line for $\mathcal{F}(A)$; that is, $\mathcal{F}(A)$ is contained in the half-plane to the left of this line.

Note also that since $e^{-\imath\theta}\mathcal{F}(e^{\imath\theta}A) = \mathcal{F}(A)$, we can use this same procedure for rotated matrices $e^{\imath\theta}A$, $\theta \in [0, 2\pi)$. If λ_θ denotes the largest eigenvalue of $H(e^{\imath\theta}A)$ and v_θ the corresponding unit eigenvector, then $v_\theta^H A v_\theta$ is a boundary point of $\mathcal{F}(A)$ and the line $\{e^{-\imath\theta}(\lambda_\theta + t\imath), \ t \in \mathbf{R}\}$ is a support line. By choosing values of θ throughout the interval $[0, 2\pi)$, the approximating polygons $p(A)$ and $P(A)$ can be made arbitrarily close to the true field of values $\mathcal{F}(A)$.

The numerical radius $\nu(A)$ also has a number of interesting properties. For any two matrices A and B, it is clear that

$$\begin{aligned} \nu(A+B) \ &= \ \max_{\|y\|=1} |y^H(A+B)y| \le \max_{\|y\|=1} |y^H Ay| + \max_{\|y\|=1} |y^H By| \\ &\le \ \nu(A) + \nu(B). \end{aligned}$$

Although the numerical radius is not itself a matrix norm (since the requirement $\nu(AB) \le \nu(A) \cdot \nu(B)$ does not always hold), it is closely related to the 2-norm:

(1.21) $$\frac{1}{2}\|A\| \le \nu(A) \le \|A\|.$$

The second inequality in (1.21) follows from the fact that for any vector y with $\|y\| = 1$, we have

$$|y^H Ay| \le \|y^H\| \cdot \|Ay\| \le \|A\|.$$

The first inequality in (1.21) is derived as follows. First note that $\nu(A) = \nu(A^H)$. Writing A in the form $A = H(A) + N(A)$, where $N(A) = (A - A^H)/2$, and noting that both $H(A)$ and $N(A)$ are normal matrices, we observe that

$$\|A\| \le \|H(A)\| + \|N(A)\| = \nu(H(A)) + \nu(N(A)).$$

Using the definition of the numerical radius this becomes

$$\begin{aligned} \|A\| \ &\le \ \frac{1}{2}\left[\max_{\|y\|=1} |y^H(A+A^H)y| + \max_{\|y\|=1} |y^H(A - A^H)y|\right] \\ &\le \ \frac{1}{2}\left[2\max_{\|y\|=1} |y^H Ay| + 2\max_{\|y\|=1} |y^H A^H y|\right] \le 2\nu(A). \end{aligned}$$

The numerical radius also satisfies the power inequality

(1.22) $$\nu(A^m) \le [\nu(A)]^m, \quad m = 1, 2, \ldots.$$

For an elementary proof, see [114] or [80, Ex. 27, p. 333].

An important property of the set of eigenvalues $\Lambda(A)$ is that if p is any polynomial, then $\Lambda(p(A)) = p(\Lambda(A))$. This can be seen from the Jordan form $A = SJS^{-1}$. If p is any polynomial then $p(A) = Sp(J)S^{-1}$ and the eigenvalues of $p(J)$ are just the diagonal elements, $p(\Lambda(A))$. Unfortunately, the field of values does not have this property: $\mathcal{F}(p(A)) \neq p(\mathcal{F}(A))$. We will see in later sections, however, that the weaker property (1.22) can still be useful in deriving error bounds for iterative methods. From (1.22) it follows that if the field of values of A is contained in a disk of radius r centered at the origin, then the field of values of A^m is contained in a disk of radius r^m centered at the origin.

There are many interesting generalizations of the field of values. One that is especially relevant to the analysis of iterative methods is as follows.

DEFINITION 1.3.7. *The* generalized field of values *of a set of matrices* $\{A_1, \ldots, A_k\}$ *in* M_n *is the subset of* $\mathbf{C}^k$ *defined by*

$$\mathcal{F}_k\left(\{A_i\}_{i=1}^k\right) = \left\{ \begin{pmatrix} y^H A_1 y \\ \vdots \\ y^H A_k y \end{pmatrix} : y \in \mathbf{C}^n, \ \|y\| = 1 \right\}.$$

Note that for $k = 1$, this is just the ordinary field of values. One can also define the *conical* generalized field of values as

$$\check{\mathcal{F}}_k\left(\{A_i\}_{i=1}^k\right) = \left\{ \begin{pmatrix} y^H A_1 y \\ \vdots \\ y^H A_k y \end{pmatrix} : y \in \mathbf{C}^n \right\}.$$

It is clear that this object is a *cone*, in the sense that if $z \in \check{\mathcal{F}}_k(\{A_i\}_{i=1}^k)$ and $\alpha > 0$, then $\alpha z \in \check{\mathcal{F}}_k(\{A_i\}_{i=1}^k)$. Note also that the conical generalized field of values is preserved by simultaneous congruence transformation: for $P \in M_n$ nonsingular, $\check{\mathcal{F}}_k(\{A_i\}_{i=1}^k) = \check{\mathcal{F}}_k(\{P^H A_i P\}_{i=1}^k)$.

Comments and Additional References.

The linear algebra facts reviewed in this chapter, along with a wealth of additional interesting material, can be found in the excellent books by Horn and Johnson [80, 81].

Part I

Krylov Subspace Approximations

Chapter 2

Some Iteration Methods

In this chapter the conjugate gradient (CG), minimal residual (MINRES), and generalized minimal residual (GMRES) algorithms are derived. These algorithms, designed for Hermitian positive definite, Hermitian indefinite, and general non-Hermitian matrices, respectively, each generate the "optimal" (in a sense, to be described later) approximation from the Krylov space (1.1).

The algorithms are first derived from other simpler iterative methods such as the method of steepest descent and Orthomin. This corresponds (roughly) to the historical development of the methods, with the optimal versions being developed as improvements upon nonoptimal algorithms. It shows how these algorithms are related to other iterative methods for solving linear systems.

Once it is recognized, however, that the goal in designing an iterative method is to generate the optimal approximation from the space (1.1), these methods can be derived from standard linear algebra techniques for locating the closest vector in a subspace to a given vector. This derivation is carried out in the latter part of section 2.4 for GMRES and in section 2.5 for MINRES and CG. This derivation has a number of advantages, such as demonstrating that the possible failure of CG for indefinite matrices corresponds to a singular tridiagonal matrix in the Lanczos algorithm and providing a basis for the derivation of other iterative techniques, such as those described in Chapter 5.

2.1. Simple Iteration.

Given a preconditioner M for the linear system $Ax = b$, a natural idea for generating approximate solutions is the following. Since a preconditioner is designed so that $M^{-1}A$ in some sense approximates the identity, $M^{-1}(b - Ax_k)$ can be expected to approximate the error $A^{-1}b - x_k$ in an approximate solution x_k. A better approximate solution x_{k+1} might therefore be obtained by taking

$$(2.1) \qquad x_{k+1} = x_k + M^{-1}(b - Ax_k).$$

This procedure of starting with an initial guess x_0 for the solution and generating successive approximations using (2.1) for $k = 0, 1, \ldots$ is sometimes called *simple iteration*, but more often it goes by different names according to the choice of M. For M equal to the diagonal of A, it is called *Jacobi* iteration;

for M equal to the lower triangle of A, it is the *Gauss–Seidel* method; for M of the form $\omega^{-1}D - L$, where D is the diagonal of A, L is the strict lower triangle of A, and ω is a relaxation parameter, it is the successive overrelaxation or *SOR* method. Preconditioners will be discussed in Part II of this book, but our concern in this section is to describe the behavior of iteration (2.1) for a given preconditioner M in terms of properties of the preconditioned matrix $M^{-1}A$.

We will see in later sections that the simple iteration procedure (2.1) can be improved upon in a number of ways. Still, it is not to be abandoned. All of these improvements require some extra work, and if the iteration matrix $M^{-1}A$ is sufficiently close to the identity, this extra work may not be necessary. Multigrid methods, which will be discussed in Chapter 12, can be thought of as very sophisticated preconditioners used with the simple iteration (2.1).

An actual implementation of (2.1) might use the following algorithm.

Algorithm 1. Simple Iteration.

Given an initial guess x_0, compute $r_0 = b - Ax_0$, and solve $M z_0 = r_0$.
For $k = 1, 2, \ldots$,

Set $x_k = x_{k-1} + z_{k-1}$.

Compute $r_k = b - Ax_k$.

Solve $M z_k = r_k$.

Let $e_k \equiv A^{-1}b - x_k$ denote the error in the approximation x_k. It follows from (2.1) that

$$(2.2) \qquad e_k = (I - M^{-1}A)e_{k-1} = \cdots = (I - M^{-1}A)^k e_0.$$

Taking norms on both sides in (2.2), we find that

$$(2.3) \qquad |||e_k||| \leq |||(I - M^{-1}A)^k||| \cdot |||e_0|||,$$

where $||| \cdot |||$ can be any vector norm provided that the matrix norm is taken to be the one *induced* by the vector norm $|||B||| \equiv \max_{|||y|||=1} |||By|||$. In this case, the bound in (2.3) is *sharp*, since, for each k, there is an initial error e_0 for which equality holds.

LEMMA 2.1.1. *The norm of the error in iteration (2.1) will approach zero and x_k will approach $A^{-1}b$ for* every *initial error e_0 if and only if*

$$\lim_{k \to \infty} |||(I - M^{-1}A)^k||| = 0.$$

Proof. It is clear from (2.3) that if $\lim_{k \to \infty} |||(I - M^{-1}A)^k||| = 0$ then $\lim_{k \to \infty} |||e_k||| = 0$. Conversely, suppose $|||(I - M^{-1}A)^k||| > \alpha > 0$ for infinitely many values of k. The vectors $e_{0,k}$ with norm 1 for which equality holds in (2.3) form a bounded infinite set in $\mathbf{C}^n$, so, by the Bolzano–Weierstrass theorem, they contain a convergent subsequence. Let e_0 be the limit of this subsequence. Then for k sufficiently large in this subsequence, we have $|||e_0 - e_{0,k}||| \le \epsilon < 1$, and

$$
\begin{aligned}
|||(I - M^1 A)^k e_0||| &\ge |||(I - M^1 A)^k e_{0,k}||| - |||(I - M^1 A)^k (e_{0,k} - e_0)||| \\
&\ge |||(I - M^1 A)^k|||(1 - \epsilon) \ge \alpha(1 - \epsilon).
\end{aligned}
$$

Since this holds for infinitely many values of k, it follows that $\lim_{k \to \infty} |||(I - M^{-1} A)^k e_0|||$, if it exists, is greater than 0. □

It was shown in section 1.3.1 that, independent of the matrix norm used in (2.3), the quantity $|||(I - M^{-1} A)^k|||^{1/k}$ approaches the *spectral radius*, $\rho(I - M^{-1} A)$ as $k \to \infty$. Thus we have the following result.

THEOREM 2.1.1. *The iteration* (2.1) *converges to* $A^{-1} b$ *for every initial error* e_0 *if and only if* $\rho(I - M^{-1} A) < 1$.

Proof. If $\rho(I - M^{-1} A) < 1$, then

$$
\lim_{k \to \infty} |||(I - M^{-1} A)^k||| = \lim_{k \to \infty} \rho(I - M^{-1} A)^k = 0,
$$

while if $\rho(I - M^{-1} A) \ge 1$, then $\lim_{k \to \infty} |||(I - M^{-1} A)^k|||$, if it exists, must be greater than or equal to 1. In either case the result then follows from Lemma 2.1.1. □

Having established necessary and sufficient conditions for convergence, we must now consider the *rate* of convergence. How many iterations will be required to obtain an approximation that is within, say, δ of the true solution? In general, this question is not so easy to answer.

Taking norms on each side in (2.2), we can write

$$
(2.4) \qquad |||e_k||| \le |||I - M^{-1} A||| \cdot |||e_{k-1}|||,
$$

from which it follows that if $|||I - M^{-1} A||| < 1$, then the error is reduced by at least this factor at each iteration. The error will satisfy $|||e_k|||/|||e_0||| \le \delta$ provided that

$$
k \ge \log \delta / \log |||I - M^{-1} A|||.
$$

It was shown in section 1.3.1 that for any $\epsilon > 0$, there is a matrix norm such that $|||I - M^{-1} A||| < \rho(I - M^{-1} A) + \epsilon$. Hence if $\rho(I - M^{-1} A) < 1$, then there is a norm for which the error is reduced *monotonically*, and convergence is at least *linear* with a reduction factor approximately equal to $\rho(I - M^{-1} A)$. Unfortunately, however, this norm is sometimes a very strange one (as might be deduced from the proof of Theorem 1.3.3, since the matrix D_t involved an exponential scaling), and it is unlikely that one would really want to measure convergence in terms of this norm!

It is usually the 2-norm or the ∞-norm or some closely related norm of the error that is of interest. For the class of *normal* matrices (diagonalizable

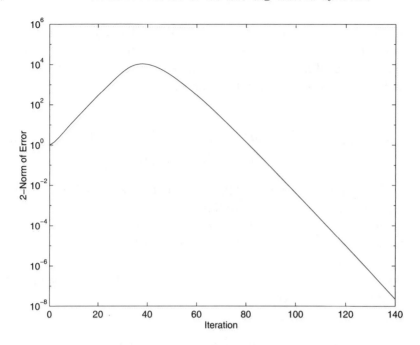

FIG. 2.1. *Simple iteration for a highly nonnormal matrix $\rho = .74$.*

matrices with a complete set of orthonormal eigenvectors), the 2-norm and the spectral radius coincide. Thus if $I - M^{-1}A$ is a normal matrix, then the 2-norm of the error is reduced by at least the factor $\rho(I - M^{-1}A)$ at each step. For nonnormal matrices, however, it is often the case that $\rho(I - M^{-1}A) < 1 < \|I - M^{-1}A\|$. In this case, the error may grow over some finite number of steps, and it is impossible to predict the number of iterations required to obtain a given level of accuracy while knowing only the spectral radius.

An example is shown in Figure 2.1. The matrix A was taken to be

$$A = \begin{pmatrix} 1 & -1.16 & & \\ .16 & \ddots & \ddots & \\ & \ddots & \ddots & -1.16 \\ & & .16 & 1 \end{pmatrix},$$

and M was taken to be the lower triangle of A. For problem size $n = 30$, the spectral radius of $I - M^{-1}A$ is about .74, while the 2-norm of $I - M^{-1}A$ is about 1.4. As Figure 2.1 shows, the 2-norm of the error increases by about four orders of magnitude over its original value before starting to decrease.

While the spectral radius generally does not determine the convergence rate of early iterations, it does describe the *asymptotic* convergence rate of (2.1). We will prove this only in the case where $M^{-1}A$ is diagonalizable and has a single eigenvalue of largest absolute value. Then, writing the eigendecomposition of $M^{-1}A$ as $V\Lambda V^{-1}$, where $\Lambda = \text{diag}(\lambda_1, \ldots, \lambda_n)$ and, say, $\rho(I - M^{-1}A) = |1 - \lambda_1| > \max_{i>1} |1 - \lambda_i|$, we have

$$V^{-1}e_k = (I - \Lambda)^k (V^{-1}e_0).$$

Assuming that the first component of $V^{-1}e_0$ is nonzero, for k sufficiently large, the largest component of $V^{-1}e_k$ will be the first, $(V^{-1}e_k)_1$. At each subsequent iteration, this dominant component is multiplied by the factor $(1 - \lambda_1)$, and so we have

$$\begin{aligned} |||e_{k+j}||| &\approx |(V^{-1}e_{k+j})_1| = \rho(I - M^{-1}A) \, |(V^{-1}e_{k+j-1})_1| \\ &\approx \rho(I - M^{-1}A) \, |||e_{k+j-1}|||, \quad j = 1, 2, \dots. \end{aligned}$$

Instead of considering the error ratios at *successive* steps as defining the asymptotic convergence rate, one might consider ratios of errors $(|||e_{k+j}|||/|||e_k|||)^{1/j}$ for any k and for j sufficiently large. Then, a more general proof that this quantity approaches the spectral radius $\rho(I - M^{-1}A)$ as $j \to \infty$ can be carried out using the Jordan form, as described in section 1.3.2. Note in Figure 2.1 that eventually the error decreases by about the factor $\rho(I - M^{-1}A) = .74$ at each step, since this matrix is diagonalizable with a single eigenvalue of largest absolute value.

2.2. Orthomin(1) and Steepest Descent.

In this section we discuss methods to improve on the simple iteration (2.1) by introducing dynamically computed parameters into the iteration. For ease of notation here and throughout the remainder of Part I of this book, we avoid explicit reference to the preconditioner M and consider A and b to be the coefficient matrix and right-hand side vector for the already preconditioned system. Sometimes we will assume that the preconditioned matrix is Hermitian. If the original coefficient matrix is Hermitian and the preconditioner M is Hermitian and *positive definite*, then one obtains a Hermitian preconditioned matrix by (implicitly) working with the modified linear system

$$L^{-1}AL^{-H}y = L^{-1}b, \quad x = L^{-H}y,$$

where $M = LL^H$ and the superscript H denotes the Hermitian transpose. If the original problem is Hermitian but the preconditioner is *indefinite*, then we consider this as a non-Hermitian problem.

One might hope to improve the iteration (2.1) by introducing a parameter a_k and setting

(2.5) $$x_{k+1} = x_k + a_k(b - Ax_k).$$

Since the residual satisfies $r_{k+1} = r_k - a_k Ar_k$, one can minimize the 2-norm of r_{k+1} by choosing

(2.6) $$a_k = \frac{\langle r_k, Ar_k \rangle}{\langle Ar_k, Ar_k \rangle}.$$

If the matrix A is Hermitian and positive definite, one might instead minimize the A-norm of the error, $\|e_{k+1}\|_A \equiv \langle e_{k+1}, Ae_{k+1}\rangle^{1/2}$. Since the error satisfies $e_{k+1} = e_k - a_k r_k$, the coefficient that minimizes this error norm is

$$(2.7) \qquad a_k = \frac{\langle e_k, Ar_k\rangle}{\langle r_k, Ar_k\rangle} = \frac{\langle r_k, r_k\rangle}{\langle r_k, Ar_k\rangle}.$$

For Hermitian positive definite problems, the iteration (2.5) with coefficient formula (2.7) is called the method of *steepest descent* because if the problem of solving the linear system is identified with that of minimizing the quadratic form $x^H A x - 2b^H x$ (which has its minimum where $Ax = b$), then the negative gradient or direction of steepest descent of this function at $x = x_k$ is $r_k = b - Ax_k$. The coefficient formula (2.6), which can be used with arbitrary nonsingular matrices A, does not have a special name like steepest descent but is a special case of a number of different methods. In particular, it can be called Orthomin(1).

By choosing a_k as in (2.6), the Orthomin(1) method produces a residual r_{k+1} that is equal to r_k minus its *projection* onto Ar_k. It follows that $\|r_{k+1}\| \leq \|r_k\|$ with equality if and only if r_k is already orthogonal to Ar_k. Recall the definition of the *field of values* $\mathcal{F}(\mathcal{B})$ of a matrix B as the set of all complex numbers of the form $y^H By / y^H y$, where y is any complex vector other than the zero vector.

THEOREM 2.2.1. *The 2-norm of the residual in iteration* (2.5) *with coefficient formula* (2.6) *decreases strictly monotonically for every initial vector r_0 if and only if $0 \notin \mathcal{F}(A^H)$.*

Proof. If $0 \in \mathcal{F}(A^H)$ and r_0 is a nonzero vector satisfying $\langle r_0, Ar_0\rangle \equiv r_0^H A^H r_0 = 0$, then $\|r_1\| = \|r_0\|$. On the other hand, if $0 \notin \mathcal{F}(A^H)$, then $\langle r_k, Ar_k\rangle$ cannot be 0 for any k and $\|r_{k+1}\| < \|r_k\|$. □

Since the field of values of A^H is just the complex conjugate of the field of values of A, the condition in the theorem can be replaced by $0 \notin \mathcal{F}(A)$.

Suppose $0 \notin \mathcal{F}(A^H)$. To show that the method (2.5–2.6) converges to the solution $A^{-1}b$, we will show not only that the 2-norm of the residual is reduced at each step but that it is reduced by at least some fixed factor, independent of k. Since the field of values is a closed set, if $0 \notin \mathcal{F}(A^H)$ then there is a positive number d—the distance of $\mathcal{F}(A^H)$ from the origin—such that $|\frac{y^H A^H y}{y^H y}| \geq d$ for all complex vectors $y \neq 0$. From (2.5) it follows that

$$r_{k+1} = r_k - a_k Ar_k,$$

and, taking the inner product of r_{k+1} with itself, we have

$$\langle r_{k+1}, r_{k+1}\rangle = \langle r_k, r_k\rangle - \frac{|\langle r_k, Ar_k\rangle|^2}{\langle Ar_k, Ar_k\rangle},$$

which can be written in the form

$$(2.8) \qquad \|r_{k+1}\|^2 = \|r_k\|^2 \left(1 - \left|\frac{r_k^H A^H r_k}{r_k^H r_k}\right|^2 \cdot \left(\frac{\|r_k\|}{\|Ar_k\|}\right)^2\right).$$

Bounding the last two factors in (2.8) independently, in terms of d and $\|A\|$, we have

$$\|r_{k+1}\|^2 \leq \|r_k\|^2 \, (1 - d^2/\|A\|^2).$$

We have proved the following theorem.

THEOREM 2.2.2. *The iteration (2.5) with coefficient formula (2.6) converges to the solution $A^{-1}b$ for all initial vectors r_0 if and only if $0 \notin \mathcal{F}(A^H)$. In this case, the 2-norm of the residual satisfies*

$$(2.9) \qquad \|r_{k+1}\| \leq \sqrt{1 - d^2/\|A\|^2} \, \|r_k\|$$

for all k, where d is the distance from the origin to the field of values of A^H.

In the special case where A is real and the Hermitian part of A, $H(A) \equiv (A + A^H)/2$, is positive definite, the distance d in (2.9) is just the smallest eigenvalue of $H(A)$. This is because the field of values of $H(A)$ is the real part of the field of values of A^H, which is convex and symmetric about the real axis and hence has its closest point to the origin on the real axis.

The bound (2.9) on the rate at which the 2-norm of the residual is reduced is not necessarily sharp, since the vectors r_k for which the first factor in (2.8) is equal to d^2 are not necessarily the ones for which the second factor is $1/\|A\|^2$. Sometimes a stronger bound can be obtained by noting that, because of the choice of a_k,

$$(2.10) \qquad \|r_{k+1}\| \leq \|I - \alpha A\| \cdot \|r_k\|$$

for *any* coefficient α. In the special case where A is Hermitian and positive definite, consider $\alpha = 2/(\lambda_n + \lambda_1)$, where λ_n is the largest and λ_1 the smallest eigenvalue of A. Inequality (2.10) then implies that

$$(2.11) \qquad \|r_{k+1}\| \leq \max_{i=1,\ldots,n} \left| 1 - \frac{2\lambda_i}{\lambda_n + \lambda_1} \right| \cdot \|r_k\| \leq \left(\frac{\kappa - 1}{\kappa + 1} \right) \|r_k\|,$$

where $\kappa = \lambda_n/\lambda_1$ is the condition number of A. (For the Hermitian positive definite case, expression (2.9) gives the significantly weaker bound $\|r_{k+1}\| \leq \sqrt{1 - \kappa^{-2}} \, \|r_k\| \approx (1 - \frac{1}{2}\kappa^{-2}) \, \|r_k\|$.)

The same argument applied to the *steepest descent* method for Hermitian positive definite problems shows that for that algorithm,

$$(2.12) \qquad \|e_{k+1}\|_A \leq \left(\frac{\kappa - 1}{\kappa + 1} \right) \|e_k\|_A.$$

In the more general non-Hermitian case, suppose the field of values of A^H is contained in a disk $\mathbf{D} = \{z \in \mathbf{C} : \; |z - \bar{c}| \leq s\}$ which does not contain the origin. Consider the choice $\alpha = 1/c$ in (2.10). It follows from (1.16–1.17) that

$$\mathcal{F}(I - (1/\bar{c})A^H) = 1 - (1/\bar{c})\mathcal{F}(A^H) \subseteq \{z \in \mathbf{C} : \; |z| \leq s/|c|\}.$$

Using relation (1.21) between the numerical radius and the norm of a matrix, we conclude that for this choice of α

$$\|I - \alpha A\| = \|I - (1/\bar{c})A^H\| \le 2\frac{s}{|c|},$$

and hence

(2.13)
$$\|r_{k+1}\| \le 2\,\frac{s}{|c|}\,\|r_k\|.$$

This estimate may be stronger or weaker than that in (2.9), depending on the exact size and shape of the field of values. For example, if $\mathcal{F}(A)$ is contained in a disk of radius $s = (\|A\| - d)/2$ centered at $c = (\|A\| + d)/2$, then (2.13) implies

$$\|r_{k+1}\| \le 2\frac{1 - d/\|A\|}{1 + d/\|A\|}\,\|r_k\|.$$

This is smaller than the bound in (2.9) if $d/\|A\|$ is greater than about .37; otherwise, (2.9) is smaller.

Stronger results have recently been proved by Eiermann [38, 39]. Suppose $\mathcal{F}(A) \subseteq \Omega$, where Ω is a compact convex set with $0 \notin \Omega$. Eiermann has shown that if $\varphi_m(z) = F_m(z)/F_m(0)$, where $F_m(z)$ is the mth-degree Faber polynomial for the set Ω (the analytic part of $(\Phi(z))^m$, where $\Phi(z)$ maps the exterior of Ω to the exterior of the unit disk), then

(2.14)
$$\nu(\varphi_m(A)) \le \max_{z\in\Omega} |\varphi_m(z)| \le c_m \min_{p_m} \max_{z\in\Omega} |p_m(z)|,$$

where the minimum is over all mth-degree polynomials with value one at the origin and the constant c_m depends on Ω but is independent of A. For $m = 1$, as in (2.10), inequality (2.14) is of limited use because the constant c_m may be larger than the inverse of the norm of the first degree minimax polynomial on Ω. We will see later, however, that for methods such as (restarted) GMRES involving higher degree polynomials, estimate (2.14) can sometimes lead to very good bounds on the norm of the residual.

Orthomin(1) can be generalized by using different inner products in (2.6). That is, if B is a Hermitian positive definite matrix, then, instead of minimizing the 2-norm of the residual, one could minimize the B-norm of the residual by taking

$$a_k = \frac{\langle r_k, BAr_k\rangle}{\langle Ar_k, BAr_k\rangle}.$$

Writing a_k in the equivalent form

$$a_k = \frac{\langle B^{1/2}r_k, (B^{1/2}AB^{-1/2})B^{1/2}r_k\rangle}{\langle B^{1/2}Ar_k, B^{1/2}Ar_k\rangle},$$

it is clear that for this variant the B-norm of the residual decreases strictly monotonically for all r_0 if and only if $0 \notin \mathcal{F}((B^{1/2}AB^{-1/2})^H)$. Using arguments similar to those used to establish Theorem 2.2.2, it can be seen that

$$(2.15) \qquad \|r_{k+1}\|_B \le \sqrt{1 - d_B^2/\|B^{1/2}AB^{-1/2}\|^2}\, \|r_k\|_B,$$

where d_B is the distance from the origin to the field of values of $(B^{1/2}AB^{-1/2})^H$. If $0 \in \mathcal{F}(A^H)$, it may still be possible to find a Hermitian positive definite matrix B such that $0 \notin \mathcal{F}((B^{1/2}AB^{-1/2})^H)$.

2.3. Orthomin(2) and CG.

The Orthomin(1) and steepest descent iterations of the previous section can be written in the form

$$x_{k+1} = x_k + a_k p_k,$$

where the *direction vector* p_k is equal to the residual r_k. The new residual and error vectors then satisfy

$$r_{k+1} = r_k - a_k A p_k, \quad e_{k+1} = e_k - a_k p_k,$$

where the coefficient a_k is chosen so that either r_{k+1} is orthogonal to Ap_k (Orthomin(1)) or, in the case of Hermitian positive definite A, so that e_{k+1} is A-orthogonal to p_k (steepest descent). Note, however, that r_{k+1} is not orthogonal to the previous vector Ap_{k-1} in the Orthomin(1) method and e_{k+1} is not A-orthogonal to the previous vector p_{k-1} in the steepest descent method.

If, in the Orthomin iteration, instead of subtracting off the projection of r_k in the direction Ar_k we subtracted off its projection in a direction like Ar_k but orthogonal to Ap_{k-1}, i.e., in the direction $A\tilde{p}_k$, where

$$\tilde{p}_k = r_k - \frac{\langle Ar_k, Ap_{k-1}\rangle}{\langle Ap_{k-1}, Ap_{k-1}\rangle} p_{k-1},$$

then we would have

$$\langle r_{k+1}, A\tilde{p}_k\rangle = 0 \text{ and } \langle r_{k+1}, Ap_{k-1}\rangle = \langle r_k, Ap_{k-1}\rangle - a_k\langle A\tilde{p}_k, Ap_{k-1}\rangle = 0.$$

Now the residual norm is minimized in the *plane* spanned by Ar_k and Ap_{k-1}, since we can write

$$r_{k+1} = r_k - a_k Ar_k + a_k b_{k-1} Ap_{k-1},$$

and the coefficients force orthogonality between r_{k+1} and $\mathrm{span}\{Ar_k, Ap_{k-1}\}$.

The new algorithm, known as Orthomin(2), is the following.

> Given an initial guess x_0, compute $r_0 = b - Ax_0$ and set $p_0 = r_0$.
> For $k = 1, 2, \ldots,$
>
> Compute Ap_{k-1}.
>
> Set $x_k = x_{k-1} + a_{k-1}p_{k-1}$, where $a_{k-1} = \frac{\langle r_{k-1}, Ap_{k-1}\rangle}{\langle Ap_{k-1}, Ap_{k-1}\rangle}$.
>
> Compute $r_k = r_{k-1} - a_{k-1}Ap_{k-1}$.
>
> Set $p_k = r_k - b_{k-1}p_{k-1}$, where $b_{k-1} = \frac{\langle Ar_k, Ap_{k-1}\rangle}{\langle Ap_{k-1}, Ap_{k-1}\rangle}$.

The new algorithm can also fail. If $\langle r_0, Ar_0 \rangle = 0$, then r_1 will be equal to r_0, and p_1 will be 0. An attempt to compute the coefficient a_1 will result in dividing 0 by 0. As for Orthomin(1), however, it can be shown that Orthomin(2) cannot fail if $0 \notin \mathcal{F}(A^H)$. If an Orthomin(2) step does succeed, then the 2-norm of the residual is reduced by at least as much as it would be reduced by an Orthomin(1) step from the same point. This is because the residual norm is minimized over a larger space—$r_k + \text{span}\{Ar_k, Ap_{k-1}\}$ instead of $r_k + \text{span}\{Ar_k\}$. It follows that the bound (2.9) holds also for Orthomin(2) when $0 \notin \mathcal{F}(A^H)$. Unfortunately, no stronger a priori bounds on the residual norm are known for Orthomin(2) applied to a general matrix whose field of values does not contain the origin although, in practice, it may perform significantly better than Orthomin(1).

In the special case when A is *Hermitian*, if the vectors at steps 1 through $k+1$ of the Orthomin(2) algorithm are defined, then r_{k+1} is minimized not just over the two-dimensional space $r_k + \text{span}\{Ar_k, Ap_{k-1}\}$ but over the $(k+1)$-dimensional space $r_0 + \text{span}\{Ap_0, \ldots, Ap_k\}$.

THEOREM 2.3.1. *Suppose that A is Hermitian, the coefficients $a_0, \ldots, a_{k-1}$ are nonzero, and the vectors $r_1, \ldots, r_{k+1}$ and $p_1, \ldots, p_{k+1}$ in the Orthomin(2) algorithm are defined. Then*

$$\langle r_{k+1}, Ap_j \rangle = \langle Ap_{k+1}, Ap_j \rangle = 0 \quad \forall j \le k.$$

It follows that of all vectors in the affine space

$$(2.16) \qquad r_0 + \text{span}\{Ar_0, A^2 r_0, \ldots, A^{k+1} r_0\},$$

r_{k+1} *has the smallest Euclidean norm. It also follows that if $a_0, \ldots, a_{n-2}$ are nonzero and $r_1, \ldots, r_n$ and $p_1, \ldots, p_n$ are defined, then $r_n = 0$.*

Proof. By construction, we have $\langle r_1, Ap_0 \rangle = \langle Ap_1, Ap_0 \rangle = 0$. Assume that $\langle r_k, Ap_j \rangle = \langle Ap_k, Ap_j \rangle = 0 \; \forall j \le k-1$. The coefficients at step $k+1$ are chosen to force $\langle r_{k+1}, Ap_k \rangle = \langle Ap_{k+1}, Ap_k \rangle = 0$. For $j \le k-1$, we have

$$\langle r_{k+1}, Ap_j \rangle = \langle r_k - a_k Ap_k, Ap_j \rangle = 0$$

by the induction hypothesis. Also,

$$
\begin{aligned}
\langle Ap_{k+1}, Ap_j \rangle &= \langle A(r_{k+1} - b_k p_k), Ap_j \rangle \\
&= \langle Ar_{k+1}, a_j^{-1}(r_j - r_{j+1}) \rangle \\
&= \bar{a}_j^{-1} \langle Ar_{k+1}, p_j + b_{j-1} p_{j-1} - p_{j+1} - b_j p_j \rangle \\
&= \bar{a}_j^{-1} \langle r_{k+1}, A(p_j + b_{j-1} p_{j-1} - p_{j+1} - b_j p_j) \rangle \\
&= 0,
\end{aligned}
$$

with the next-to-last equality holding because $A = A^H$. It is justified to write a_j^{-1} since, by assumption, $a_j \ne 0$.

It is easily checked by induction that r_{k+1} lies in the space (2.16) and that $\text{span}\{Ap_0, \ldots, Ap_k\} = \text{span}\{Ar_0, \ldots, A^{k+1} r_0\}$. Since r_{k+1} is orthogonal to span$\{Ap_0, \ldots, Ap_k\}$, it follows that r_{k+1} is the vector in the space (2.16) with minimal Euclidean norm. For $k = n - 1$, this implies that $r_n = 0$. □

The assumption in Theorem 2.3.1 that $r_1, \ldots, r_{k+1}$ and $p_1, \ldots, p_{k+1}$ are defined is actually implied by the other hypothesis. It can be shown that these vectors are defined provided that $a_0, \ldots, a_{k-1}$ are defined and nonzero and r_k is nonzero.

An algorithm that approximates the solution of a Hermitian linear system $Ax = b$ by minimizing the residual over the affine space (2.16) is known as the *MINRES* algorithm. It should not be implemented in the form given here, however, unless the matrix is positive (or negative) definite because, as noted, this iteration can fail if $0 \in \mathcal{F}(A)$. An appropriate implementation of the MINRES algorithm for Hermitian indefinite linear systems is derived in section 2.5.

In a similar way, the steepest descent method for Hermitian positive definite matrices can be modified so that it eliminates the A-projection of the error in a direction that is already A-orthogonal to the previous direction vector, i.e., in the direction

$$\tilde{p}_k = r_k - \frac{\langle r_k, Ap_{k-1} \rangle}{\langle p_{k-1}, Ap_{k-1} \rangle} p_{k-1}.$$

Then we have

$$\langle e_{k+1}, A\tilde{p}_k \rangle = \langle e_{k+1}, Ap_{k-1} \rangle = 0,$$

and the A-norm of the error is minimized over the two-dimensional affine space $e_k + \text{span}\{r_k, p_{k-1}\}$. The algorithm that does this is called the *CG* method. It is usually implemented with slightly different (but equivalent) coefficient formulas, as shown in Algorithm 2.

Algorithm 2. Conjugate Gradient Method (CG)
(for Hermitian positive definite problems)

Given an initial guess x_0, compute $r_0 = b - Ax_0$ and set $p_0 = r_0$.
For $k = 1, 2, \ldots,$

Compute Ap_{k-1}.

Set $x_k = x_{k-1} + a_{k-1}p_{k-1}$, where $a_{k-1} = \frac{\langle r_{k-1}, r_{k-1} \rangle}{\langle p_{k-1}, Ap_{k-1} \rangle}$.

Compute $r_k = r_{k-1} - a_{k-1}Ap_{k-1}$.

Set $p_k = r_k + b_{k-1}p_{k-1}$, where $b_{k-1} = \frac{\langle r_k, r_k \rangle}{\langle r_{k-1}, r_{k-1} \rangle}$.

It is left as an exercise for the reader to prove that these coefficient formulas are equivalent to the more obvious expressions

$$(2.17) \qquad a_{k-1} = \frac{\langle r_{k-1}, p_{k-1} \rangle}{\langle p_{k-1}, Ap_{k-1} \rangle}, \quad b_{k-1} = -\frac{\langle r_k, Ap_{k-1} \rangle}{\langle p_{k-1}, Ap_{k-1} \rangle}.$$

Since the CG algorithm is used only with positive definite matrices, the coefficients are always defined, and it can be shown, analogous to the MINRES method, that the A-norm of the error is actually minimized over the much larger affine space $e_0 + \text{span}\{p_0, p_1, \ldots, p_k\}$.

THEOREM 2.3.2. *Assume that A is Hermitian and positive definite. The CG algorithm generates the exact solution to the linear system $Ax = b$ in at most n steps. The error, residual, and direction vectors generated before the exact solution is obtained are well defined and satisfy*

$$\langle e_{k+1}, Ap_j \rangle = \langle p_{k+1}, Ap_j \rangle = \langle r_{k+1}, r_j \rangle = 0 \quad \forall j \leq k.$$

It follows that of all vectors in the affine space

$$(2.18) \qquad e_0 + \text{span}\{Ae_0, A^2 e_0, \ldots, A^{k+1} e_0\},$$

e_{k+1} *has the smallest A-norm.*

Proof. Since A is positive definite, it is clear that the coefficients in the CG algorithm are well defined unless a residual vector is zero, in which case the exact solution has been found. Assume that $r_0, \ldots, r_k$ are nonzero. By the choice of a_0, it is clear that $\langle r_1, r_0 \rangle = \langle e_1, Ap_0 \rangle = 0$, and from the choice of b_0 it follows that

$$
\begin{aligned}
\langle p_1, Ap_0 \rangle &= \langle r_1, Ap_0 \rangle + \frac{\langle r_1, r_1 \rangle}{\langle r_0, r_0 \rangle} \langle p_0, Ap_0 \rangle \\
&= \langle r_1, a_0^{-1}(r_0 - r_1) \rangle + a_0^{-1} \langle r_1, r_1 \rangle = 0,
\end{aligned}
$$

where the last equality holds because $\langle r_1, r_0 \rangle = 0$ and a_0^{-1} is real. Assume that

$$\langle e_k, Ap_j \rangle = \langle p_k, Ap_j \rangle = \langle r_k, r_j \rangle = 0 \quad \forall j \leq k-1.$$

Then we also have

$$\langle p_k, Ap_k \rangle = \langle r_k + b_{k-1}p_{k-1}, Ap_k \rangle = \langle r_k, Ap_k \rangle,$$

$$\langle r_k, p_k \rangle = \langle r_k, r_k + b_{k-1}p_{k-1} \rangle = \langle r_k, r_k \rangle,$$

so, by the choice of a_k, it follows that

$$\langle r_{k+1}, r_k \rangle = \langle r_k, r_k \rangle - a_k \langle r_k, Ap_k \rangle = \langle r_k, r_k \rangle - \langle r_k, r_k \rangle = 0,$$

$$\langle e_{k+1}, Ap_k \rangle = \langle r_{k+1}, p_k \rangle = \langle r_k, p_k \rangle - a_k \langle Ap_k, p_k \rangle = \langle r_k, r_k \rangle - \langle r_k, r_k \rangle = 0.$$

From the choice of b_k, we have

$$\langle p_{k+1}, Ap_k \rangle = \langle r_{k+1}, Ap_k \rangle + \frac{\langle r_{k+1}, r_{k+1} \rangle}{\langle r_k, r_k \rangle} \langle p_k, Ap_k \rangle$$

$$= \langle r_{k+1}, a_k^{-1}(r_k - r_{k+1}) \rangle + a_k^{-1} \langle r_{k+1}, r_{k+1} \rangle = 0.$$

For $j \leq k - 1$, we have

$$\langle e_{k+1}, Ap_j \rangle = \langle e_k - a_k p_k, Ap_j \rangle = 0,$$

$$\langle r_{k+1}, r_j \rangle = \langle r_k - a_k Ap_k, r_j \rangle = -a_k \langle p_k, A(p_j - b_{j-1}p_{j-1}) \rangle = 0,$$

$$\langle p_{k+1}, Ap_j \rangle = \langle r_{k+1} + b_k p_k, Ap_j \rangle = \langle r_{k+1}, a_j^{-1}(r_j - r_{j+1}) \rangle = 0,$$

so, by induction, the desired equalities are established.

It is easily checked by induction that e_{k+1} lies in the space (2.18) and that span$\{p_0, \ldots, p_k\}$ = span$\{Ae_0, \ldots, A^{k+1}e_0\}$. Since e_{k+1} is A-orthogonal to span$\{p_0, \ldots, p_k\}$, it follows that e_{k+1} is the vector in the space (2.18) with minimal A-norm, and it also follows that if the exact solution is not obtained before step n, then $e_n = 0$. $\square$

2.4. Orthodir, MINRES, and GMRES.

Returning to the case of general matrices A, the idea of minimizing over a larger subspace can be extended, at the price of having to save and orthogonalize against additional vectors at each step. To minimize the 2-norm of the residual r_{k+1} over the j-dimensional affine space

$$r_k + \text{span}\{Ap_k, Ap_{k-1}, \ldots, Ap_{k-j+1}\} = r_k + \text{span}\{Ar_k, Ap_{k-1}, \ldots, Ap_{k-j+1}\},$$

set

$$(2.19) \qquad p_k = r_k - \sum_{\ell=1}^{j-1} b_{k-\ell}^{(k)} p_{k-\ell}, \quad b_{k-\ell}^{(k)} = \frac{\langle Ar_k, Ap_{k-\ell} \rangle}{\langle Ap_{k-\ell}, Ap_{k-\ell} \rangle}.$$

This defines the Orthomin(j) procedure. Unfortunately, the algorithm can still fail if $0 \in \mathcal{F}(A^H)$, and again the only proven a priori bound on the residual norm is estimate (2.9), although this bound is often pessimistic.

It turns out that the possibility of failure can be eliminated by replacing r_k in formula (2.19) with Ap_{k-1}. This algorithm, known as *Orthodir*, generally has worse convergence behavior than Orthomin for $j < n$, however. The bound (2.9) can no longer be established because the space over which the norm of r_{k+1} is minimized may not contain the vector Ar_k.

An exception is the case of *Hermitian* matrices, where it can be shown that for $j = 3$, the Orthodir(j) algorithm minimizes the 2-norm of the residual over the entire affine space

$$(2.20) \quad r_0 + \text{span}\{Ap_0, Ap_1, \ldots, Ap_k\} = r_0 + \text{span}\{Ar_0, A^2 r_0, \ldots, A^k r_0\}.$$

This provides a reasonable implementation of the MINRES algorithm described in section 2.3.

> Given an initial guess x_0, compute $r_0 = b - Ax_0$ and set $p_0 = r_0$.
> Compute $s_0 = Ap_0$. For $k = 1, 2, \ldots,$

Set $x_k = x_{k-1} + a_{k-1}p_{k-1}$, where $a_{k-1} = \frac{\langle r_{k-1}, s_{k-1}\rangle}{\langle s_{k-1}, s_{k-1}\rangle}$.

Compute $r_k = r_{k-1} - a_{k-1}s_{k-1}$.

Set $p_k = s_{k-1}$, $s_k = As_{k-1}$. For $\ell = 1, 2$,

$$b_{k-\ell}^{(k)} = \frac{\langle s_k, s_{k-\ell}\rangle}{\langle s_{k-\ell}, s_{k-\ell}\rangle},$$

$$p_k \longleftarrow p_k - b_{k-\ell}^{(k)}p_{k-\ell},$$

$$s_k \longleftarrow s_k - b_{k-\ell}^{(k)}s_{k-\ell}.$$

A difficulty with this algorithm is that in finite precision arithmetic, the vectors s_k, which are supposed to be equal to Ap_k, may differ from this if there is much cancellation in the computation of s_k. This could be corrected with an occasional extra matrix–vector multiplication to explicitly set $s_k = Ap_k$ at the end of an iteration. Another possible implementation is given in section 2.5.

For general non-Hermitian matrices, if $j = n$, then the Orthodir(n) algorithm minimizes the 2-norm of the residual at each step k over the affine space in (2.20). It follows that the exact solution is obtained in n or fewer steps (assuming exact arithmetic) but at the cost of storing up to n search directions p_k (as well as auxiliary vectors $s_k \equiv Ap_k$) and orthogonalizing against k direction vectors at each step $k = 1, \ldots, n$. If the full n steps are required, then Orthodir(n) requires $O(n^2)$ storage and $O(n^3)$ work, just as would be required by a standard dense Gaussian elimination routine. The power of the method lies in the fact that at each step the residual norm is minimized over the space (2.20) so that, hopefully, an acceptably good approximate solution can be obtained in far fewer than n steps.

There is another way to compute the approximation x_k for which the norm of r_k is minimized over the space (2.20). This method requires about half the storage of Orthodir(n) (no auxiliary vectors) and has better numerical properties. It is the *GMRES method*.

The GMRES method uses the modified Gram–Schmidt process to construct an orthonormal basis for the Krylov space span$\{r_0, Ar_0, \ldots, A^k r_0\}$. When the modified Gram–Schmidt process is applied to this space in the form given below it is called *Arnoldi's method*.

Arnoldi Algorithm.

Given q_1 with $\|q_1\| = 1$. For $j = 1, 2, \ldots$,

$\tilde{q}_{j+1} = Aq_j$. For $i = 1, \ldots, j$, $\quad h_{ij} = \langle \tilde{q}_{j+1}, q_i\rangle$, $\quad \tilde{q}_{j+1} \longleftarrow \tilde{q}_{j+1} - h_{ij}q_i$.

$h_{j+1,j} = \|\tilde{q}_{j+1}\|$, $\quad q_{j+1} = \tilde{q}_{j+1}/h_{j+1,j}$.

If Q_k is the n-by-k matrix with the orthonormal basis vectors $q_1, \ldots, q_k$ as columns, then the Arnoldi iteration can be written in matrix form as

$$(2.21) \qquad AQ_k = Q_k H_k + h_{k+1,k}q_{k+1}\xi_k^T = Q_{k+1}H_{k+1,k}.$$

Here H_k is the k-by-k upper Hessenberg matrix with (i,j)-element equal to $h_{i,j}$ for $j = 1, \ldots, k$, $i = 1, \ldots, \min\{j+1, k\}$, and all other elements zero. The vector ξ_k is the kth unit k-vector, $(0, \ldots, 0, 1)^T$. The $k+1$-by-k matrix $H_{k+1,k}$ is the matrix whose top k-by-k block is H_k and whose last row is zero except for the $(k+1, k)$-element, which is $h_{k+1,k}$. Pictorially, the matrix equation (2.21) looks like

In the GMRES method, the approximate solution x_k is taken to be of the form $x_k = x_0 + Q_k y_k$ for some vector y_k; that is, x_k is x_0 plus some linear combination of the orthonormal basis vectors for the Krylov space. To obtain the approximation for which $r_k = r_0 - A Q_k y_k$ has a minimal 2-norm, the vector y_k must solve the least squares problem

$$\min_y \|r_0 - A Q_k y\| = \min_y \|r_0 - Q_{k+1} H_{k+1,k} y\|$$

$$= \min_y \|Q_{k+1}(\beta \xi_1 - H_{k+1,k} y)\| = \min_y \|\beta \xi_1 - H_{k+1,k} y\|,$$

where $\beta = \|r_0\|$, ξ_1 is the first unit $(k+1)$-vector $(1, 0 \ldots, 0)^T$, and the second equality is obtained by using the fact that $Q_{k+1}\xi_1$, the first orthonormal basis vector, is just r_0/β.

The basic steps of the GMRES algorithm are as follows.

Given x_0, compute $r_0 = b - A x_0$ and set $q_1 = r_0/\|r_0\|$. For $k = 1, 2, \ldots$,

Compute q_{k+1} and $h_{i,k}$, $i = 1, \ldots, k+1$ using the Arnoldi algorithm.

Form $x_k = x_0 + Q_k y_k$, where y_k is the solution to the least squares problem $\min_y \|\beta \xi_1 - H_{k+1,k} y\|$.

A standard method for solving the least squares problem $\min_y \|\beta \xi_1 - H_{k+1,k} y\|$ is to factor the $k+1$-by-k matrix $H_{k+1,k}$ into the product of a $k+1$-by-$k+1$ unitary matrix F^H and a $k+1$-by-k upper triangular matrix R (that is, the top k-by-k block is upper triangular and the last row is 0). This factorization, known as the QR factorization, can be accomplished using plane rotations. The solution y_k is then obtained by solving the upper triangular system

(2.22) $R_{k \times k}\, y = \beta\, (F \xi_1)_{k \times 1},$

where $R_{k \times k}$ is the top k-by-k block of R and $(F \xi_1)_{k \times 1}$ is the top k entries of the first column of F.

Given the QR factorization of $H_{k+1,k}$, we would like to be able to compute the QR factorization of the next matrix $H_{k+2,k+1}$ with as little work as possible. To see how this can be done, let F_i denote the rotation matrix that rotates the unit vectors ξ_i and ξ_{i+1} through the angle θ_i:

$$F_i = \begin{pmatrix} I & & & \\ & c_i & s_i & \\ & -\bar{s}_i & c_i & \\ & & & I \end{pmatrix},$$

where $c_i \equiv \cos(\theta_i)$ and $s_i \equiv \sin(\theta_i)$. The dimension of the matrix F_i, that is, the size of the second identity block, will depend on the context in which it is used. Assume that the rotations F_i, $i = 1, \ldots, k$ have previously been applied to $H_{k+1,k}$ so that

$$(F_k F_{k-1} \cdots F_1) \, H_{k+1,k} = R^{(k)} = \begin{pmatrix} x & x & \cdots & x \\ & x & \cdots & x \\ & & \ddots & \vdots \\ & & & x \\ 0 & 0 & \cdots & 0 \end{pmatrix},$$

where the x's denote nonzeros. In order to obtain $R^{(k+1)}$, the upper triangular factor for $H_{k+2,k+1}$, first premultiply the last column of $H_{k+2,k+1}$ by the previous rotations to obtain

$$(F_k F_{k-1} \cdots F_1) \, H_{k+2,k+1} = \begin{pmatrix} x & x & \cdots & x & x \\ & x & \cdots & x & x \\ & & \ddots & \vdots & \vdots \\ & & & x & x \\ 0 & 0 & \cdots & 0 & d \\ 0 & 0 & \cdots & 0 & h \end{pmatrix},$$

where the $(k+2, k+1)$-entry, h, is just $h_{k+2,k+1}$, since this entry is unaffected by the previous rotations. The next rotation, F_{k+1}, is chosen to eliminate this entry by setting $c_{k+1} = |d|/\sqrt{|d|^2 + |h|^2}$, $\bar{s}_{k+1} = c_{k+1}h/d$ if $d \neq 0$, and $c_{k+1} = 0$, $s_{k+1} = 1$ if $d = 0$. Note that the $(k+1)$st diagonal entry of $R^{(k+1)}$ is nonzero since h is nonzero (assuming the exact solution to the linear system has not already been computed), and, if $d \neq 0$, then this diagonal element is $(d/|d|)\sqrt{|d|^2 + |h|^2}$ while if $d = 0$, the diagonal element is h.

The right-hand side in (2.22) is computed by applying each of the rotations $F_1, \ldots, F_k$ to the unit vector ξ_1. The absolute value of the last entry of this $(k+1)$-vector, multiplied by β, is the 2-norm of the residual at step k since

$$\|b - Ax_k\| = \|\beta \xi_1 - F^H R y_k\| = \|\beta F \xi_1 - R y_k\|,$$

and $\beta F \xi_1 - R y_k$ is zero except for its bottom entry, which is just the bottom entry of $\beta F \xi_1$.

The GMRES algorithm can be written in the following form.

Algorithm 3. Generalized Minimal Residual Algorithm (GMRES).

Given x_0, compute $r_0 = b - Ax_0$ and set $q_1 = r_0/\|r_0\|$.
Initialize $\xi = (1, 0, \ldots, 0)^T$, $\beta = \|r_0\|$. For $k = 1, 2, \ldots$,

Compute q_{k+1} and $h_{i,k} \equiv H(i, k)$, $i = 1, \ldots, k + 1$, using the Arnoldi algorithm.

Apply $F_1, \ldots, F_{k-1}$ to the last column of H; that is,
  For $i = 1, \ldots, k - 1$,

$$\begin{pmatrix} H(i,k) \\ H(i+1,k) \end{pmatrix} \leftarrow \begin{pmatrix} c_i & s_i \\ -\bar{s}_i & c_i \end{pmatrix} \begin{pmatrix} H(i,k) \\ H(i+1,k) \end{pmatrix}$$

Compute the kth rotation, c_k and s_k, to annihilate the $(k + 1, k)$ entry of H.[1]

Apply kth rotation to ξ and to last column of H:

$$\begin{pmatrix} \xi(k) \\ \xi(k+1) \end{pmatrix} \leftarrow \begin{pmatrix} c_k & s_k \\ -\bar{s}_k & c_k \end{pmatrix} \begin{pmatrix} \xi(k) \\ 0 \end{pmatrix}$$

$$H(k,k) \leftarrow c_k H(k,k) + s_k H(k+1,k), \quad H(k+1,k) \leftarrow 0.$$

If residual norm estimate $\beta|\xi(k + 1)|$ is sufficiently small, then
  Solve upper triangular system $H_{k \times k}\, y_k = \beta\, \xi_{k \times 1}$.
  Compute $x_k = x_0 + Q_k y_k$.

The (full) GMRES algorithm described above may be impractical because of increasing storage and work requirements, if the number of iterations needed to solve the linear system is large. The GMRES(j) algorithm is defined by simply restarting GMRES every j steps, using the latest iterate as the initial guess for the next GMRES cycle. In Chapter 3, we discuss the convergence rate of full and restarted GMRES.

2.5. Derivation of MINRES and CG from the Lanczos Algorithm.

When the matrix A is Hermitian, the Arnoldi algorithm of the previous section can be simplified to a 3-term recurrence known as the Lanczos algorithm. Slightly different (but mathematically equivalent) coefficient formulas are normally used in the Hermitian case.

---

[1]The formula is $c_k = |H(k,k)|/\sqrt{|H(k,k)|^2 + |H(k+1,k)|^2}$, $\bar{s}_k = c_k H(k+1,k)/H(k,k)$, but a more robust implementation should be used. See, for example, BLAS routine DROTG [32].

Lanczos Algorithm (for Hermitian matrices A).

Given q_1 with $\|q_1\| = 1$, set $\beta_0 = 0$. For $j = 1, 2, \ldots,$

$\tilde{q}_{j+1} = Aq_j - \beta_{j-1}q_{j-1}$. Set $\alpha_j = \langle \tilde{q}_{j+1}, q_j \rangle, \quad \tilde{q}_{j+1} \longleftarrow \tilde{q}_{j+1} - \alpha_j q_j.$

$\beta_j = \|\tilde{q}_{j+1}\|, \quad q_{j+1} = \tilde{q}_{j+1}/\beta_j.$

To see that the vectors constructed by this algorithm are the same as those constructed by the Arnoldi algorithm when the matrix A is Hermitian, we must show that they form an orthonormal basis for the Krylov space formed from A and q_1. It is clear that the vectors lie in this Krylov space and each vector has norm one because of the choice of the β_j's. From the formula for α_j, it follows that $\langle q_{j+1}, q_j \rangle = 0$. Suppose $\langle q_k, q_i \rangle = 0$ for $i \neq k$ whenever $k, i \leq j$. Then

$$
\begin{aligned}
\langle \tilde{q}_{j+1}, q_{j-1} \rangle &= \langle Aq_j - \alpha_j q_j - \beta_{j-1}q_{j-1}, q_{j-1} \rangle = \langle Aq_j, q_{j-1} \rangle - \beta_{j-1} \\
&= \langle q_j, Aq_{j-1} \rangle - \beta_{j-1} \\
&= \langle q_j, \tilde{q}_j + \alpha_{j-1}q_{j-1} + \beta_{j-2}q_{j-2} \rangle - \beta_{j-1} = \langle q_j, \tilde{q}_j \rangle - \beta_{j-1} = 0.
\end{aligned}
$$

For $i < j - 1$, we have

$$
\begin{aligned}
\langle \tilde{q}_{j+1}, q_i \rangle &= \langle Aq_j - \alpha_j q_j - \beta_{j-1}q_{j-1}, q_i \rangle = \langle Aq_j, q_i \rangle \\
&= \langle q_j, Aq_i \rangle \\
&= \langle q_j, \tilde{q}_{i+1} + \alpha_i q_i + \beta_{i-1}q_{i-1} \rangle = 0.
\end{aligned}
$$

Thus the vectors $q_1, \ldots, q_{j+1}$ form an orthonormal basis for the Krylov space span$\{q_1, Aq_1, \ldots, A^j q_1\}$.

The Lanczos algorithm can be written in matrix form as

$$(2.23) \qquad AQ_k = Q_k T_k + \beta_k q_{k+1} \xi_k^T = Q_{k+1} T_{k+1,k},$$

where Q_k is the n-by-k matrix whose columns are the orthonormal basis vectors $q_1, \ldots, q_k$, ξ_k is the kth unit k-vector, and T_k is the k-by-k Hermitian tridiagonal matrix of recurrence coefficients:

$$(2.24) \qquad T_k = \begin{pmatrix} \alpha_1 & \beta_1 & & \\ \beta_1 & \ddots & \ddots & \\ & \ddots & \ddots & \beta_{k-1} \\ & & \beta_{k-1} & \alpha_k \end{pmatrix}.$$

The $k+1$-by-k matrix $T_{k+1,k}$ has T_k as its upper k-by-k block and $\beta_k \xi_k^T$ as its last row.

It was shown in section 2.3 that the MINRES and CG algorithms generate the Krylov space approximations x_k for which the 2-norm of the residual and the A-norm of the error, respectively, are minimal. That is, if q_1 in the

Lanczos algorithm is taken to be r_0/β, $\beta = \|r_0\|$, then the algorithms generate approximate solutions of the form $x_k = x_0 + Q_k y_k$, where y_k is chosen to minimize the appropriate error norm. For the MINRES algorithm, y_k solves the least squares problem

$$
\begin{aligned}
\min_y \|r_0 - AQ_k y\| &= \min_y \|r_0 - Q_{k+1} T_{k+1,k} y\| \\
&= \min_y \|Q_{k+1}(\beta \xi_1 - T_{k+1,k} y)\| \\
&= \min_y \|\beta \xi_1 - T_{k+1,k} y\|,
\end{aligned}
$$

(2.25)

similar to the GMRES algorithm of the previous section.

For the MINRES algorithm, however, there is no need to save the orthonormal basis vectors generated by the Lanczos algorithm. Hence a different formula is needed to compute the approximate solution x_k. Let $R_{k \times k}$ be the upper k-by-k block of the triangular factor R in the QR decomposition of $T_{k+1,k} = F^H R$, as described in the previous section. Since $T_{k+1,k}$ is tridiagonal, $R_{k \times k}$ has only three nonzero diagonals. Define $P_k \equiv (p_0, \ldots, p_{k-1}) \equiv Q_k R_{k \times k}^{-1}$. Then p_0 is a multiple of q_1 and successive columns of P_k can be computed using the fact that $P_k R_{k \times k} = Q_k$:

$$
p_{k-1} = \left(q_k - b_{k-2}^{(k-1)} p_{k-2} - b_{k-3}^{(k-1)} p_{k-3} \right) / b_{k-1}^{(k-1)},
$$

where $b_{k-\ell}^{(k-1)}$ is the $(k - \ell + 1, k)$-entry of $R_{k \times k}$. Recall from the arguments of the previous section that $b_{k-1}^{(k-1)}$ is nonzero, provided that the exact solution to the linear system has not been obtained already. The approximate solution x_k can be updated from x_{k-1} since

$$
x_k = x_0 + P_k \beta \, (F\xi_1)_{k \times 1} = x_{k-1} + a_{k-1} p_{k-1},
$$

where a_{k-1} is the kth entry of $\beta(F\xi_1)$. This leads to the following implementation of the MINRES algorithm.

Algorithm 4. Minimal Residual Algorithm (MINRES)
(for Hermitian Problems).

Given x_0, compute $r_0 = b - Ax_0$ and set $q_1 = r_0/\|r_0\|$.
Initialize $\xi = (1, 0, \ldots, 0)^T$, $\beta = \|r_0\|$. For $k = 1, 2, \ldots,$

Compute q_{k+1}, $\alpha_k \equiv T(k, k)$, and $\beta_k \equiv T(k+1, k) \equiv T(k, k+1)$
using the Lanczos algorithm.

Apply F_{k-2} and F_{k-1} to the last column of T; that is,

$$\begin{pmatrix} T(k-2, k) \\ T(k-1, k) \end{pmatrix} \leftarrow \begin{pmatrix} c_{k-2} & s_{k-2} \\ -\bar{s}_{k-2} & c_{k-2} \end{pmatrix} \begin{pmatrix} 0 \\ T(k-1, k) \end{pmatrix}, \quad \text{if } k > 2,$$

$$\begin{pmatrix} T(k-1, k) \\ T(k, k) \end{pmatrix} \leftarrow \begin{pmatrix} c_{k-1} & s_{k-1} \\ -\bar{s}_{k-1} & c_{k-1} \end{pmatrix} \begin{pmatrix} T(k-1, k) \\ T(k, k) \end{pmatrix}, \quad \text{if } k > 1.$$

Compute the kth rotation, c_k and s_k, to annihilate the $(k+1, k)$-entry of T.[2]

Apply kth rotation to ξ and to last column of T:

$$\begin{pmatrix} \xi(k) \\ \xi(k+1) \end{pmatrix} \leftarrow \begin{pmatrix} c_k & s_k \\ -\bar{s}_k & c_k \end{pmatrix} \begin{pmatrix} \xi(k) \\ 0 \end{pmatrix}.$$

$$T(k, k) \leftarrow c_k T(k, k) + s_k T(k+1, k), \quad T(k+1, k) \leftarrow 0.$$

Compute $p_{k-1} = [q_k - T(k-1, k)p_{k-2} - T(k-2, k)p_{k-3}]/T(k, k)$,
where undefined terms are zero for $k \leq 2$.

Set $x_k = x_{k-1} + a_{k-1}p_{k-1}$, where $a_{k-1} = \beta\xi(k)$.

For the CG method, y_k is chosen to make the residual r_k orthogonal to the columns of Q_k. For positive definite matrices A, this minimizes the A-norm of the error since $e_k = e_0 - Q_k y_k$ has minimal A-norm when it is A-orthogonal to the columns of Q_k, i.e., when $Q_k^H A e_k = Q_k^H r_k = 0$. Note that the criterion that r_k be orthogonal to the columns of Q_k can be enforced for Hermitian *indefinite* problems as well, although it does not correspond to minimizing any obvious error norm. The vector y_k for the CG algorithm satisfies

$$(2.26) \qquad Q_k^H(r_0 - AQ_k y_k) = \beta\xi_1 - T_k y_k = 0;$$

that is, y_k is the solution to the k-by-k linear system $T_k y = \beta\xi_1$. While the least squares problem (2.25) always has a solution, the linear system (2.26) has a unique solution if and only if T_k is *nonsingular*. When A is positive definite, it follows from the minimax characterization of eigenvalues of a Hermitian

---

[2]The formula is $c_k = |T(k, k)|/\sqrt{|T(k, k)|^2 + |T(k+1, k)|^2}$, $\bar{s}_k = c_k T(k+1, k)/T(k, k)$, but a more robust implementation should be used. See, for example, BLAS routine DROTG [32].

matrix that the tridiagonal matrix $T_k = Q_k^H A Q_k$ is also positive definite, since its eigenvalues lie between the smallest and largest eigenvalues of A.

If T_k is positive definite, and sometimes even if it is not, then one way to solve (2.26) is to factor T_k in the form

$$(2.27) \qquad\qquad T_k = L_k D_k L_k^H,$$

where L_k is a unit lower bidiagonal matrix and D_k is a diagonal matrix. One would like to be able to compute not only y_k but the approximate solution $x_k = x_0 + Q_k y_k$ without saving all of the basis vectors $q_1, \ldots, q_k$. The factorization (2.27) can be updated easily from one step to the next, since L_k and D_k are the k-by-k principal submatrices of L_{k+1} and D_{k+1}. If we define $P_k \equiv (p_0, \ldots p_{k-1}) \equiv Q_k L_k^{-H}$, then the columns of P_k are A-orthogonal since

$$P_k^H A P_k = L_k^{-1} Q_k^H A Q_k L_k^{-H} = L_k^{-1} T_k L_k^{-H} = D_k,$$

and since P_k satisfies $P_k L_k^H = Q_k$, the columns of P_k can be computed in succession via the recurrence

$$p_k = q_k - \bar{b}_{k-1} p_{k-1},$$

where b_{k-1} is the $(k, k-1)$-entry of L_k. It is not difficult to see that the columns of P_k are, to within constant factors, the direction vectors from the CG algorithm of section 2.3. The Lanczos vectors are normalized versions of the CG residuals, with opposite signs at every other step. With this factorization, then, x_k is given by

$$x_k = x_0 + Q_k T_k^{-1} \beta e_1 = x_0 + P_k D_k^{-1} L_k^{-1} \beta e_1,$$

and since x_{k-1} satisfies

$$x_{k-1} = x_0 + P_{k-1} D_{k-1}^{-1} L_{k-1}^{-1} \beta e_1,$$

it can be seen that x_k satisfies

$$x_k = x_{k-1} + a_{k-1} p_{k-1},$$

where $a_{k-1} = d_k^{-1} \beta (L_k^{-1})_{k,1}$ and d_k is the (k, k)-entry of D_k. The coefficient a_{k-1} is defined, provided that L_k is invertible and $d_k \neq 0$.

With this interpretation it can be seen that if the CG algorithm of section 2.3 were applied to a Hermitian *indefinite* matrix, then it would fail at step k if and only if the LDL^H factorization of T_k does not exist. If this factorization exists for $T_1, \ldots, T_{k-1}$, then it can fail to exist at step k only if T_k is singular. For indefinite problems, it is possible that T_k will be singular, but subsequent tridiagonal matrices, e.g., T_{k+1}, will be nonsingular. The CG algorithm of section 2.3 cannot recover from a singular intermediate matrix T_k. To overcome this difficulty, Paige and Saunders [111] proposed a CG-like algorithm based on the LQ-factorization of the tridiagonal matrices. This algorithm is known as SYMMLQ. We will not derive the SYMMLQ algorithm here, because it does not have any nice error minimization property like the MINRES method for Hermitian indefinite problems. We will show in Chapter 5, however, that the residuals in the two methods are closely related.

Comments and Additional References.

For a discussion of simple iteration methods (e.g., Jacobi, Gauss–Seidel, SOR), the classic books of Varga [135] and Young [144] are still highly recommended.

The CG algorithm was originally proposed by Hestenes and Stiefel [79] and appeared in a related work at the same time by Lanczos [90]. The Orthomin method described in this chapter was first introduced by Vinsome [137], and the Orthodir algorithm was developed by Young and Jea [145]. Subsequently, Saad and Schultz invented the GMRES algorithm [119]. Theorem 2.2.2 was proved in the special case when the Hermitian part of the matrix is positive definite by Eisenstat, Elman, and Schultz [40].

The MINRES implementation described in section 2.5 was given by Paige and Saunders [111], as was the first identification of the CG algorithm with the Lanczos process followed by LDL^H-factorization of the tridiagonal matrix. The idea of minimizing the 2-norm of the residual for Hermitian indefinite problems was contained in the original Hestenes and Stiefel paper, implemented in the form we have referred to here as Orthomin(2). This is sometimes called the *conjugate residual* method. For an implementation that uses the Orthomin(2) algorithm when it is safe to do so and switches to Orthodir(3) in other circumstances, see Chandra [25].

A useful bibliography of work on the CG and Lanczos algorithms between 1948 and 1976 is contained in [59].

Exercises.

2.1. Use the Jordan form discussed in section 1.3.2 to describe the asymptotic convergence of simple iteration for a nondiagonalizable matrix.

2.2. Show that if A and b are *real* and $0 \in \mathcal{F}(A^H)$, then there is a *real* initial vector for which the Orthomin iteration fails.

2.3. Give an example of a problem for which Orthomin(1) converges but simple iteration (with no preconditioner) does not. Give an example of a problem for which simple iteration converges but Orthomin(j) does not for any $j < n$.

2.4. Verify that the coefficient formulas in (2.17) are equivalent to those in Algorithm 2.

2.6. Express the entries of the tridiagonal matrix generated by the Lanczos algorithm in terms of the coefficients in the CG algorithm (Algorithm 2). (Hint: Write down the 3-term recurrence for $q_{k+1} \equiv (-1)^k r_k/\|r_k\|$ in terms of q_k and q_{k-1}.)

2.7. Derive a necessary and sufficient condition for the convergence of the restarted GMRES algorithm, GMRES(j), in terms of the *generalized field of values* defined in section 1.3.3; that is, show that GMRES(j) converges to the solution of the linear system for all initial vectors if and only if the zero vector is not contained in the set

$$
\mathcal{F}_j(A, A^2, \ldots, A^j) = \left\{ \begin{pmatrix} y^H Ay \\ y^H A^2 y \\ \vdots \\ y^H A^j y \end{pmatrix} : y \in C^n, \ \|y\| = 1 \right\}.
$$

2.8. Show that for a *normal* matrix whose eigenvalues have real parts greater than or equal to the imaginary parts in absolute value, the GMRES(2) iteration converges for all initial vectors. (Hint: Since $r_2 = P_2(A)r_0$ for a certain second-degree polynomial P_2 with $P_2(0) = 1$ and since P_2 minimizes the 2-norm of the residual over all such polynomials, we have $\|r_2\| \leq \min_{p_2} \|p_2(A)\| \cdot \|r_0\|$. If a second-degree polynomial p_2 with value 1 at the origin can be found which satisfies $\|p_2(A)\| < 1$, then this will show that each GMRES(2) cycle reduces the residual norm by at least a constant factor. Since A is normal, its eigendecomposition can be written in the form $A = U\Lambda U^T$, where U is unitary and $\Lambda = \mathrm{diag}(\lambda_1, \ldots, \lambda_n)$; it follows that $\|p_2(A)\| = \max_{i=1,\ldots,n} |p_2(\lambda_i)|$. Hence, find a polynomial $p_2(z)$ that is strictly less than 1 in absolute value throughout a region

$$
\{z : |\mathrm{Re}(z)| \geq |\mathrm{Im}(z)|, \ |z| \leq \max_{i=1,\ldots,n} |\lambda_i|\},
$$

containing the spectrum of A.)

Chapter 3

Error Bounds for CG, MINRES, and GMRES

It was shown in Chapter 2 that the CG, MINRES, and GMRES algorithms each generate the optimal approximate solution from a Krylov subspace, where "optimal" is taken to mean having an error with minimal A-norm in the case of CG or having a residual with minimal 2-norm in the case of MINRES and GMRES. In this chapter we derive bounds on the appropriate error norm for the optimal approximation from a Krylov subspace.

A goal is to derive a *sharp* upper bound on the reduction in the A-norm of the error for CG or in the 2-norm of the residual for both MINRES and GMRES—that is, an upper bound that is independent of the initial vector but that is actually attained for certain initial vectors. This describes the *worst-case* behavior of the algorithms (for a given matrix A). It can sometimes be shown that the "typical" behavior of the algorithms is not much different from the worst-case behavior. That is, if the initial vector is random, then convergence may be only moderately faster than for the worst initial vector. For certain special initial vectors, however, convergence may be much faster than the worst-case analysis would suggest. Still, it is usually the same analysis that enables one to identify these "special" initial vectors, and it is often clear how the bounds must be modified to account for special properties of the initial vector.

For *normal* matrices, a sharp upper bound on the appropriate error norm is known. This is not the case for nonnormal matrices, and a number of possible approaches to this problem are discussed in section 3.2.

3.1. Hermitian Problems—CG and MINRES.

It was shown in Chapter 2 that the A-norm of the error in the CG algorithm for Hermitian positive definite problems and the 2-norm of the residual in the MINRES algorithm for general Hermitian problems are minimized over the spaces

$$e_0 + \text{span}\{Ae_0, A^2e_0, \ldots, A^ke_0\} \quad \text{and} \quad r_0 + \text{span}\{Ar_0, A^2r_0, \ldots, A^kr_0\},$$

respectively. It follows that the CG error vector and the MINRES residual vector at step k can be written in the form

$$(3.1) \qquad e_k = P_k^C(A)e_0, \quad r_k = P_k^M(A)r_0,$$

where P_k^C and P_k^M are kth-degree polynomials with value 1 at the origin and, of all such polynomials that could be substituted in (3.1), P_k^C gives the error of minimal A-norm in the CG algorithm and P_k^M gives the residual of minimal 2-norm in the MINRES algorithm. In other words, the error e_k in the CG approximation satisfies

$$(3.2) \qquad \|e_k\|_A = \min_{p_k} \|p_k(A)e_0\|_A$$

and the residual r_k in the MINRES algorithm satisfies

$$(3.3) \qquad \|r_k\| = \min_{p_k} \|p_k(A)r_0\|,$$

where the minimum is taken over all polynomials p_k of degree k or less with $p_k(0) = 1$.

In this section we derive bounds on the expressions in the right-hand sides of (3.2) and (3.3) that are *independent* of the direction of the initial error e_0 or residual r_0, although they do depend on the size of these quantities. A sharp upper bound is derived involving all of the *eigenvalues* of A, and then simpler (but nonsharp) bounds are given based on knowledge of just a few of the eigenvalues of A.

Let an eigendecomposition of A be written as $A = U \Lambda U^H$, where U is a unitary matrix and $\Lambda = \mathrm{diag}(\lambda_1, \ldots, \lambda_n)$ is a diagonal matrix of eigenvalues. If A is positive definite, define $A^{1/2}$ to be $U\Lambda^{1/2}U^H$. Then the A-norm of a vector v is just the 2-norm of the vector $A^{1/2}v$. Equalities (3.2) and (3.3) imply that

$$\|e_k\|_A = \min_{p_k} \|A^{1/2}p_k(A)e_0\| = \min_{p_k} \|Up_k(\Lambda)U^H A^{1/2}e_0\|$$
$$(3.4) \qquad \leq \min_{p_k} \|p_k(\Lambda)\| \cdot \|e_0\|_A,$$

$$(3.5) \qquad \|r_k\| = \min_{p_k} \|Up_k(\Lambda)U^H r_0\| \leq \min_{p_k} \|p_k(\Lambda)\| \cdot \|r_0\|,$$

with the inequalities following, because if $\hat{p}_k$ is the polynomial that minimizes $\|p_k(\Lambda)\|$, then

$$\min_{p_k} \|Up_k(\Lambda)U^H w\| \leq \|U\hat{p}_k(\Lambda)U^H w\| \leq \|U\hat{p}_k(\Lambda)U^H\| \, \|w\| = \|\hat{p}_k(\Lambda)\| \, \|w\|$$

for any vector w. Of course, the polynomial that minimizes the expressions in the equalities of (3.4) and (3.5) is *not* necessarily the same one that minimizes $\|p_k(\Lambda)\|$ in the inequalities. The MINRES and CG polynomials depend on the initial vector, while this polynomial does not. Hence it is not immediately obvious that the bounds in (3.4) and (3.5) are *sharp*, that is, that they can actually be attained for certain initial vectors. It turns out that this is the case, however. See, for example, [63, 68, 85]. For each k there is an initial vector e_0 for which the CG polynomial at step k is the polynomial that minimizes $\|p_k(\Lambda)\|$ and for which equality holds in (3.4). An analogous result holds for MINRES.

The sharp upper bounds (3.4) and (3.5) can be written in the form

$$(3.6) \qquad \|e_k\|_A/\|e_0\|_A \leq \min_{p_k} \max_{i=1,\ldots,n} |p_k(\lambda_i)| \quad \text{for CG,}$$

$$(3.7) \qquad \|r_k\|/\|r_0\| \leq \min_{p_k} \max_{i=1,\ldots,n} |p_k(\lambda_i)| \quad \text{for MINRES.}$$

The problem of describing the convergence of these algorithms therefore reduces to one in approximation theory—how well can one approximate zero on the set of eigenvalues of A using a kth-degree polynomial with value 1 at the origin? While there is no simple expression for the maximum value of the minimax polynomial on a discrete set of points, this minimax polynomial can be calculated if the eigenvalues of A are known; more importantly, this sharp upper bound provides intuition as to what constitutes "good" and "bad" eigenvalue distributions. Eigenvalues tightly clustered around a single point (away from the origin) are good, for instance, because the polynomial $(1 - z/c)^k$ is small in absolute value at all points near c. Widely spread eigenvalues, especially if they lie on both sides of the origin, are bad, because a low-degree polynomial with value 1 at the origin cannot be small at a large number of such points.

Since one usually has only limited information about the eigenvalues of A, it is useful to have error bounds that involve only a few properties of the eigenvalues. For example, in the CG algorithm for Hermitian positive definite problems, knowing only the largest and smallest eigenvalues of A, one can obtain an error bound by considering the minimax polynomial on the *interval* from λ_{min} to λ_{max}, i.e., the Chebyshev polynomial shifted to the interval and scaled to have value 1 at the origin.

THEOREM 3.1.1. *Let e_k be the error at step k of the CG algorithm applied to the Hermitian positive definite linear system $Ax = b$. Then*

$$(3.8) \qquad \frac{\|e_k\|_A}{\|e_0\|_A} \leq 2 \left[\left(\frac{\sqrt{\kappa} - 1}{\sqrt{\kappa} + 1} \right)^k + \left(\frac{\sqrt{\kappa} + 1}{\sqrt{\kappa} - 1} \right)^k \right]^{-1} \leq 2 \left(\frac{\sqrt{\kappa} - 1}{\sqrt{\kappa} + 1} \right)^k,$$

where $\kappa = \lambda_{max}/\lambda_{min}$ is the ratio of the largest to smallest eigenvalue of A.

Proof. Consider the kth scaled and shifted Chebyshev polynomial on the interval $[\lambda_{min}, \lambda_{max}]$

$$(3.9) \qquad p_k(z) = T_k \left(\frac{2z - \lambda_{max} - \lambda_{min}}{\lambda_{max} - \lambda_{min}} \right) \Big/ T_k \left(\frac{-\lambda_{max} - \lambda_{min}}{\lambda_{max} - \lambda_{min}} \right),$$

where $T_k(z)$ is the Chebyshev polynomial of the first kind on the interval $[-1, 1]$ satisfying

$$T_0(z) = 1, \quad T_1(z) = z,$$

$$T_{j+1}(z) = 2zT_j(z) - T_{j-1}(z), \quad j = 1, 2, \ldots.$$

In the interval $[-1, 1]$, we have $T_k(z) = \cos(k\cos^{-1}(z))$, so $|T_k(z)| \leq 1$ and the absolute value of the numerator in (3.9) is bounded by 1 for z in the interval $[\lambda_{min}, \lambda_{max}]$. It attains this bound at the endpoints of the interval and at $k-1$ interior points. To determine the size of the denominator in (3.9), note that outside the interval $[-1, 1]$, we have

$$T_k(z) = \cosh(k\cosh^{-1}z),$$

so if z is of the form $z = \cosh(\ln y) = \frac{1}{2}(y + y^{-1})$, then $T_k(z) = \frac{1}{2}(y^k + y^{-k})$. The argument in the denominator of (3.9) can be expressed in the form $\frac{1}{2}(y + y^{-1})$ if y satisfies

$$-\frac{\lambda_{max} + \lambda_{min}}{\lambda_{max} - \lambda_{min}} = -\frac{\kappa + 1}{\kappa - 1} = \frac{1}{2}(y + y^{-1}),$$

which is equivalent to the quadratic equation

$$\frac{1}{2}y^2 + \frac{\kappa + 1}{\kappa - 1}y + \frac{1}{2} = 0.$$

Solving this equation, we find

$$y = -\frac{\sqrt{\kappa} + 1}{\sqrt{\kappa} - 1} \quad \text{or} \quad y = -\frac{\sqrt{\kappa} - 1}{\sqrt{\kappa} + 1}.$$

In either case, the denominator in (3.9) has absolute value equal to

$$\frac{1}{2}\left[\left(\frac{\sqrt{\kappa} + 1}{\sqrt{\kappa} - 1}\right)^k + \left(\frac{\sqrt{\kappa} - 1}{\sqrt{\kappa} + 1}\right)^k\right],$$

and from this the result (3.8) follows. □

Knowing only the largest and smallest eigenvalues of a Hermitian positive definite matrix A, bound (3.8) is the best possible. If the interior eigenvalues of A lie at the points where the Chebyshev polynomial p_k in (3.9) attains its maximum absolute value on $[\lambda_{min}, \lambda_{max}]$, then for a certain initial error e_0, the CG polynomial will be equal to the Chebyshev polynomial, and the bound in (3.8) will actually be attained at step k.

If additional information is available about the interior eigenvalues of A, one can often improve on the estimate (3.8) while maintaining a simpler expression than the sharp bound (3.6). Suppose, for example, that A has one eigenvalue much larger than the others, say, $\lambda_1 \leq \cdots \leq \lambda_{n-1} << \lambda_n$, that is, $\lambda_n/\lambda_{n-1} >> 1$. Consider a polynomial p_k that is the product of a linear factor that is zero at λ_n and the $(k-1)$st-degree scaled and shifted Chebyshev polynomial on the interval $[\lambda_1, \lambda_{n-1}]$:

$$p_k(z) = \left[T_{k-1}\left(\frac{2z - \lambda_{n-1} - \lambda_1}{\lambda_{n-1} - \lambda_1}\right)\middle/T_{k-1}\left(\frac{-\lambda_{n-1} - \lambda_1}{\lambda_{n-1} - \lambda_1}\right)\right] \cdot \left(\frac{\lambda_n - z}{\lambda_n}\right).$$

Since the second factor is zero at λ_n and less than one in absolute value at each of the other eigenvalues, the maximum absolute value of this polynomial on $\{\lambda_1, \ldots, \lambda_n\}$ is less than the maximum absolute value of the first factor on $\{\lambda_1, \ldots, \lambda_{n-1}\}$. Using arguments like those in Theorem 3.1.1, it follows that

$$(3.10) \qquad \frac{\|e_k\|_A}{\|e_0\|_A} \leq 2 \left(\frac{\sqrt{\kappa_{n-1}} - 1}{\sqrt{\kappa_{n-1}} + 1} \right)^{k-1}, \qquad \kappa_{n-1} = \frac{\lambda_{n-1}}{\lambda_1}.$$

Similarly, if the matrix A has just a few large outlying eigenvalues, say, $\lambda_1 \leq \cdots \leq \lambda_{n-\ell} << \lambda_{n-\ell+1} \leq \cdots \leq \lambda_n$ (i.e., $\lambda_{n-\ell+1}/\lambda_{n-\ell} >> 1$), one can consider a polynomial p_k that is the product of an ℓth-degree factor that is zero at each of the outliers (and less than one in magnitude at each of the other eigenvalues) and a scaled and shifted Chebyshev polynomial of degree $k - \ell$ on the interval $[\lambda_1, \lambda_{n-\ell}]$. Bounding the size of this polynomial gives

$$(3.11) \qquad \frac{\|e_k\|_A}{\|e_0\|_A} \leq 2 \left(\frac{\sqrt{\kappa_{n-\ell}} - 1}{\sqrt{\kappa_{n-\ell}} + 1} \right)^{k-\ell}, \qquad \kappa_{n-\ell} = \frac{\lambda_{n-\ell}}{\lambda_1}.$$

Analogous results hold for the 2-norm of the residual in the MINRES algorithm applied to a Hermitian positive definite linear system, and the proofs are identical. For example, for any $\ell \geq 0$, we have

$$(3.12) \qquad \frac{\|r_k\|}{\|r_0\|} \leq 2 \left(\frac{\sqrt{\kappa_{n-\ell}} - 1}{\sqrt{\kappa_{n-\ell}} + 1} \right)^{k-\ell}, \qquad \kappa_{n-\ell} = \frac{\lambda_{n-\ell}}{\lambda_1} \quad \text{(MINRES)}.$$

For Hermitian *indefinite* problems, a different polynomial must be considered. We derive only a simple estimate in the case when the eigenvalues of A are contained in two intervals $[a, b] \bigcup [c, d]$, where $a < b < 0 < c < d$ and $b - a = d - c$. In this case, the kth-degree polynomial with value 1 at the origin that has minimal maximum deviation from 0 on $[a, b] \bigcup [c, d]$ is given by

$$(3.13) \qquad p_k(z) = T_\ell(q(z))/T_\ell(q(0)), \qquad q(z) = 1 + \frac{2(z - b)(z - c)}{ad - bc},$$

where $\ell = [\frac{k}{2}]$, $[\cdot]$ denotes the integer part, and T_ℓ is the ℓth Chebyshev polynomial. Note that the function $q(z)$ maps each of the intervals $[a, b]$ and $[c, d]$ to the interval $[-1, 1]$. It follows that for $z \in [a, b] \bigcup [c, d]$, the absolute value of the numerator in (3.13) is bounded by 1. The size of the denominator is determined in the same way as before: if $q(0) = \frac{1}{2}(y + y^{-1})$, then $T_\ell(q(0)) = \frac{1}{2}(y^\ell + y^{-\ell})$. To determine y, we must solve the equation

$$q(0) = \frac{ad + bc}{ad - bc} = \frac{1}{2}(y + y^{-1})$$

or the quadratic equation

$$\frac{1}{2}y^2 - \frac{ad + bc}{ad - bc}y + \frac{1}{2} = 0.$$

This equation has the solutions

$$y = \frac{\sqrt{|ad|} - \sqrt{|bc|}}{\sqrt{|ad|} + \sqrt{|bc|}} \quad \text{or} \quad \frac{\sqrt{|ad|} + \sqrt{|bc|}}{\sqrt{|ad|} - \sqrt{|bc|}}.$$

It follows that the norm of the kth MINRES residual is bounded by

$$(3.14) \qquad \frac{\|r_k\|}{\|r_0\|} \leq 2 \left(\frac{\sqrt{|ad|} - \sqrt{|bc|}}{\sqrt{|ad|} + \sqrt{|bc|}} \right)^{[k/2]}.$$

In the special case when $a = -d$ and $b = -c$ (so the two intervals are placed symmetrically about the origin), the bound in (3.14) becomes

$$(3.15) \qquad \frac{\|r_k\|}{\|r_0\|} \leq 2 \left(\frac{d/c - 1}{d/c + 1} \right)^{[k/2]}.$$

This is the bound one would obtain at step $[k/2]$ for a Hermitian positive definite matrix with condition number $(d/c)^2$! It is as difficult to approximate zero on two intervals situated symmetrically about the origin as it is to approximate zero on a single interval lying on one side of the origin whose ratio of largest to smallest point is equal to the *square* of that in the 2-interval problem. Remember, however, that estimate (3.14) implies better approximation properties for intervals not symmetrically placed about the origin. For further discussion of approximation problems on two intervals, see [50, secs. 3.3–3.4]

3.2. Non-Hermitian Problems—GMRES.

Like MINRES for Hermitian problems, the GMRES algorithm for general linear systems produces a residual at step k whose 2-norm satisfies (3.3). To derive a bound on the expression in (3.3) that is independent of the direction of r_0, we could proceed as in the previous section by employing an eigendecomposition of A. To this end, assume that A is diagonalizable and let $A = V \Lambda V^{-1}$ be an eigendecomposition, where $\Lambda = \text{diag}(\lambda_1, \ldots, \lambda_n)$ is a diagonal matrix of eigenvalues and the columns of V are right eigenvectors of A (normalized in any desired way). Then it follows from (3.3) that

$$\|r_k\| = \min_{p_k} \|V p_k(\Lambda) V^{-1} r_0\| \leq \kappa(V) \min_{p_k} \|p_k(\Lambda)\| \cdot \|r_0\|,$$

$$(3.16) \qquad \|r_k\|/\|r_0\| \leq \kappa(V) \min_{p_k} \max_{i=1,\ldots,n} |p_k(\lambda_i)|,$$

where $\kappa(V) = \|V\| \cdot \|V^{-1}\|$ is the condition number of the eigenvector matrix V. We will assume that the columns of V have been scaled to make this condition number as small as possible. As in the Hermitian case, the polynomial that minimizes $\|V p_k(\Lambda) V^{-1} r_0\|$ is not necessarily the one that minimizes $\|p_k(\Lambda)\|$, and it is not clear whether the bound in (3.16) is sharp. It turns out that if A is a *normal* matrix (a diagonalizable matrix with a complete set of orthonormal eigenvectors), then $\kappa(V) = 1$ and the bound in (3.16) is sharp [68, 85]. In this case, as in the Hermitian case, the problem of describing the convergence of GMRES reduces to a problem in approximation theory—how well can one approximate zero on the set of *complex* eigenvalues using a kth-degree polynomial with value 1 at the origin? We do not have simple estimates, such as that obtained in Theorem 3.1.1 based on the ratio of largest to smallest eigenvalue, but one's intuition about good and bad eigenvalue distributions in the complex plane still applies. Eigenvalues tightly clustered about a single point (away from the origin) are good, since the polynomial $(1 - z/c)^k$ is small at all points close to c in the complex plane. Eigenvalues all around the origin are bad because (by the maximum principle) it is impossible to have a polynomial that is 1 at the origin and less than 1 everywhere on some closed curve around the origin. Similarly, a low-degree polynomial cannot be 1 at the origin and small in absolute value at many points distributed all around the origin.

If the matrix A is nonnormal but has a fairly well-conditioned eigenvector matrix V, then the bound (3.16), while not necessarily sharp, gives a reasonable estimate of the actual size of the residual. In this case again, it is A's eigenvalue distribution that essentially determines the behavior of GMRES.

In general, however, the behavior of GMRES cannot be determined from eigenvalues alone. In fact, it is shown in [72, 69] that *any* nonincreasing curve represents a plot of residual norm versus iteration number for the GMRES method applied to some problem; moreover, that problem can be taken to have any desired eigenvalues. Thus, for example, eigenvalues tightly clustered around 1 are not necessarily good for nonnormal matrices, as they are for normal ones.

A simple way to see this is to consider a matrix A with the following sparsity pattern:

$$(3.17) \qquad \begin{pmatrix} 0 & * & 0 & \cdots & 0 \\ 0 & * & * & \cdots & 0 \\ \vdots & \vdots & \vdots & \ddots & \\ 0 & * & * & \cdots & * \\ * & * & * & \cdots & * \end{pmatrix},$$

where the $*$'s represent any values and the other entries are 0. If the initial residual r_0 is a multiple of the first unit vector $\xi_1 \equiv (1, 0, \ldots, 0)^T$, then Ar_0 is a multiple of ξ_n, $A^2 r_0$ is a linear combination of ξ_n and ξ_{n-1}, etc. All vectors $A^k r_0$, $k = 1, \ldots, n-1$ are orthogonal to r_0, so the optimal approximation from the space $x_0 + \mathrm{span}\{r_0, Ar_0, \ldots, A^{k-1} r_0\}$ is simply $x_k = x_0$ for $k = 1, \ldots, n-1$; i.e., GMRES makes no progress until step n! Now, the class of matrices of the form (3.17) includes, for example, all *companion matrices*:

$$\begin{pmatrix} 0 & 1 & & \\ 0 & 0 & \ddots & \\ \vdots & \vdots & & 1 \\ c_0 & c_1 & \cdots & c_{n-1} \end{pmatrix}$$

The eigenvalues of this matrix are the roots of the polynomial $z^n - \sum_{j=0}^{n-1} c_j z^j$, and the coefficients $c_0, \ldots, c_{n-1}$ can be chosen to make this matrix have any desired eigenvalues. If GMRES is applied to (3.17) with a different initial residual, say, a random r_0, then, while some progress will be made before step n, it is likely that a significant residual component will remain, until that final step.

Of course, one probably would not use the GMRES algorithm to solve a linear system with the sparsity pattern of that in (3.17), but the same result holds for any matrix that is unitarily similar to one of the form (3.17). Note that (3.17) is simply a permuted lower triangular matrix. Every matrix is unitarily similar to a lower triangular matrix, but, fortunately, most matrices are *not* unitarily similar to one of the form (3.17)!

When the eigenvector matrix V is extremely ill-conditioned, the bound (3.16) is less useful. It may be greater than 1 for all $k < n$, but we know from other arguments that $\|r_k\|/\|r_0\| \leq 1$ for all k. In such cases, it is not clear whether GMRES converges poorly or whether the bound (3.16) is simply a large overestimate of the actual residual norm. Attempts have been made to delineate those cases in which GMRES actually does converge poorly from those for which GMRES converges well and the bound (3.16) is just a large overestimate.

Different bounds on the residual norm can be obtained based on the field of values of A, provided $0 \notin \mathcal{F}(A)$. For example, suppose $\mathcal{F}(A)$ is contained in a disk $\mathbf{D} = \{z \in \mathbf{C} : |z - c| \leq s\}$ which does not contain the origin. Consider the polynomial $p_k(z) = (1 - z/c)^k$. It follows from (1.16–1.17) that

$$\mathcal{F}(I - (1/c)A) = 1 - (1/c)\mathcal{F}(A) \subseteq \{z \in \mathbf{C} : |z| \leq s/|c|\}$$

and hence that $\nu(I - (1/c)A) \leq s/|c|$. The power inequality (1.22) implies that $\nu((I - (1/c)A)^k) \leq (s/|c|)^k$ and hence, by (1.21),

$$\|p_k(A)\| \leq 2 \left(\frac{s}{|c|}\right)^k.$$

It follows that the GMRES residual norm satisfies

(3.18) $$\|r_k\|/\|r_0\| \leq 2 \left(\frac{s}{|c|}\right)^k,$$

and this bound holds for the restarted GMRES algorithm GMRES(j) provided that $j \geq k$. It is somewhat stronger than the bound (2.13) for GMRES(1) (which is the same as Orthomin(1)), because the factor 2 does not have to be raised to the kth power. Still, (3.18) sometimes gives a significant overestimate of the actual GMRES residual norm. In many cases, a disk $\mathbf{D}$ may have to be much larger than $\mathcal{F}(A)$ in order to include $\mathcal{F}(A)$.

Using the recent result (2.14), however, involving the Faber polynomials for an arbitrary convex set S that contains $\mathcal{F}(A)$ and not the origin, one can more closely fit $\mathcal{F}(A)$ while choosing S so that the Faber polynomials for S are close to the minimax polynomials. For example, suppose $\mathcal{F}(A)$ is contained in the ellipse

$$E_s(\gamma, \delta) = \{z \in \mathbf{C} : \quad |z - (\delta - \gamma)| + |z - (\delta + \gamma)| \leq |\gamma|(s + s^{-1})\},$$

with foci $\delta \pm \gamma$ and semi-axes $|\gamma|(s \pm s^{-1})$. Assume that $0 \notin E_s(\gamma, \delta)$. The kth Faber polynomial for E_s is just the kth Chebyshev polynomial of the first kind translated to the interval $[\delta - \gamma, \delta + \gamma]$. When this polynomial is normalized to have value one at the origin, its maximum value on E_s can be shown to be

$$|F_k(z)/F_k(0)| \leq (s^k + s^{-k}) \frac{\kappa^k}{1 - \kappa^{2k}}, \quad z \in E_s,$$

where

$$\kappa = \left| \frac{\delta - \sqrt{\delta^2 - \gamma^2}}{\gamma} \right|$$

and the branch of the square root is chosen so that $\kappa < 1$. We assume here that $s < \kappa^{-1}$. For further details, see [38, 39].

Inequality (2.14) still does not lead to a sharp bound on the residual norm in most cases, and it can be applied only when $0 \notin \mathcal{F}(A)$. Another approach to estimating $\|p(A)\|$ in terms of the size of $p(z)$ in some region of the complex plane has been suggested by Trefethen [129]. It is the idea of *pseudo-eigenvalues*.

For any polynomial p, the matrix $p(A)$ can be written as a Cauchy integral

$$(3.19) \qquad p(A) = \frac{1}{2\pi i} \int_{\Gamma} p(z)(zI - A)^{-1} dz,$$

where Γ is any simple closed curve or union of simple closed curves containing the spectrum of A. Taking norms on each side in (3.19) and replacing the norm of the integral by the length $\mathcal{L}(\Gamma)$ of the curve times the maximum norm of the integrand gives

$$(3.20) \qquad \|p(A)\| \leq \frac{\mathcal{L}(\Gamma)}{2\pi} \max_{z \in \Gamma} \|p(z)(zI - A)^{-1}\|.$$

Now, if we consider a curve Γ_ϵ on which the resolvent norm $\|(zI - A)^{-1}\|$ is constant, say, $\|(zI - A)^{-1}\| = \epsilon^{-1}$, then (3.20) implies

$$(3.21) \qquad \|p(A)\| \leq \frac{\mathcal{L}(\Gamma_\epsilon)}{2\pi\epsilon} \max_{z \in \Gamma_\epsilon} |p(z)|.$$

The curve on which $\|(zI - A)^{-1}\| = \epsilon^{-1}$ is referred to as the boundary of the ϵ-pseudospectrum of A:

$$\Lambda_\epsilon \equiv \{z : \|(zI - A)^{-1}\| \geq \epsilon^{-1}\}.$$

From (3.21) and the optimality of the GMRES approximation, it follows that the GMRES residual r_k satisfies

$$(3.22) \qquad \|r_k\|/\|r_0\| \leq \frac{\mathcal{L}(\Gamma_\epsilon)}{2\pi\epsilon} \min_{p_k} \max_{z \in \Gamma_\epsilon} |p_k(z)|$$

for any choice of the parameter ϵ. For certain problems, and with carefully chosen values of ϵ, the bound (3.22) may be much smaller than that in (3.16). Still, the bound (3.22) is not sharp, and for some problems there is no choice of ϵ that yields a realistic estimate of the actual GMRES residual [72]. It is easy to see where the main overestimate occurs. In going from (3.19) to (3.20) and replacing the norm of the integral by the length of the curve times the maximum norm of the integrand, one may lose important cancellation properties of the integral.

Each of the inequalities (3.16), (3.18), and (3.22) provides bounds on the GMRES residual by bounding the quantity $\min_{p_k} \|p_k(A)\|$. Now, the worst-case behavior of GMRES is given by

$$(3.23) \qquad \|r_k\| = \max_{\|r_0\|=1} \min_{p_k} \|p_k(A)r_0\|.$$

The polynomial p_k depends on r_0. Until recently, it was an open question whether the right-hand side of (3.23) was equal to the quantity

$$(3.24) \qquad \min_{p_k} \|p_k(A)\| \equiv \min_{p_k} \max_{\|r_0\|=1} \|p_k(A)r_0\|.$$

It is known that the right-hand sides of (3.23) and (3.24) are equal if A is a normal matrix or if the dimension of A is less than or equal to 3 or if $k = 1$, and many numerical experiments have shown that these two quantities are equal (to within the accuracy limits of the computation) for a wide variety of matrices and values of k. Recently, however, it has been shown that the two quantities may differ. Faber et al. [44] constructed an example in which the right-hand side of (3.24) is 1, while that of (3.23) is .9995. Subsequently, Toh [127] generated examples in which the ratio of the right-hand side of (3.24) to that of (3.23) can be made arbitrarily large by varying a parameter in the matrix. Thus neither of the approaches leading to inequalities (3.16) and (3.22) can be expected to yield a sharp bound on the size of the GMRES residual, and it remains an open problem to describe the convergence of GMRES in terms of some simple characteristic properties of the coefficient matrix.

Exercises.

3.1. Suppose a positive definite matrix has a small, well-separated eigenvalue, $\lambda_1 \ll \lambda_2 \leq \cdots \leq \lambda_n$ (that is, $\lambda_1/\lambda_2 \ll 1$). Derive an error bound for

CG or MINRES using the maximum value of a polynomial that is the product of a linear factor that is 0 at λ_1 and a $(k-1)$st-degree Chebyshev polynomial on the interval $[\lambda_2, \lambda_n]$. Is it more advantageous to have a small, well-separated eigenvalue or a large, well-separated eigenvalue as in (3.10)? (For the derivation of many other such error bounds, see [132].)

3.2. Consider the 4-by-4 matrix

$$A = \begin{pmatrix} 1 & \epsilon & & \\ -1 & 1/\epsilon & & \\ & 1 & \epsilon & \\ & & & -1 \end{pmatrix}, \quad \epsilon > 0.$$

This is the example devised by Toh [127] to demonstrate the difference between expressions (3.23) and (3.24).

(a) Show that the polynomial of degree 3 or less with value one at the origin that minimizes $\|p(A)\|$ over all such polynomials is

$$p_*(z) = 1 - \frac{3}{5}z^2$$

and that $\|p_*(A)\| = \frac{4}{5}$, independent of ϵ. (Hint: First show that $p_*(z)$ must be even by using the uniqueness of p_* and the fact that A^T is unitarily similar to $-A$ via the matrix

$$Q = \begin{pmatrix} & & & -1 \\ & & 1 & \\ & -1 & & \\ 1 & & & \end{pmatrix},$$

which implies that $\|p(-A)\| = \|p(A^T)\| = \|p(A)\|$ for any polynomial p. Now consider polynomials p_γ of the form $1 + \gamma z^2$ for various scalars γ. Determine the singular values of $p_\gamma(A)$ analytically to show that

$$\sigma_{max}^2(\gamma) = \frac{1}{2}\left(2(\gamma+1)^2 + \gamma^2 + |\gamma|\sqrt{4(\gamma+1)^2 + \gamma^2}\right).$$

Differentiate with respect to γ and set the derivative to zero to show that $\gamma = \frac{3}{5}$ minimizes σ_{max}.)

(b) Show that for any vector b with $\|b\| = 1$ there is a polynomial p_b of degree 3 or less with $p_b(0) = 1$ such that

$$\|p_b(A)b\| \leq 4\sqrt{\epsilon} + 5\epsilon.$$

(Hint: First note that if $b = (b_1, b_2, b_3, b_4)^T$ then

$$Ab = \begin{pmatrix} b_1 \\ -b_2 + b_3/\epsilon \\ b_3 \\ -b_4 \end{pmatrix} + \epsilon \begin{pmatrix} b_2 \\ 0 \\ b_4 \\ 0 \end{pmatrix}, \quad A^2 b = b + \begin{pmatrix} b_3 \\ b_4 \\ 0 \\ 0 \end{pmatrix},$$

$$A^3 b = Ab + \begin{pmatrix} b_3 \\ -b_4 \\ 0 \\ 0 \end{pmatrix} + \epsilon \begin{pmatrix} b_4 \\ 0 \\ 0 \\ 0 \end{pmatrix}.$$

If $|b_3| \geq \sqrt{\epsilon}$, take

$$p_b(z) = 1 - z - z^2 + \left(\frac{2b_4}{b_3} + 1 \right) z^3,$$

and if $|b_3| < \sqrt{\epsilon}$ take

$$p_b(z) = 1 + z - z^2 + (\epsilon - 1)z^3.)$$

Chapter 4

Effects of Finite Precision Arithmetic

In the previous chapter, error bounds were derived for the CG, MINRES, and GMRES algorithms, using the fact that these methods find the *optimal* approximation from a Krylov subspace. In the Arnoldi algorithm, on which the GMRES method is based, all of the Krylov space basis vectors are retained, and a new vector is formed by explicitly orthogonalizing against all previous vectors using the modified Gram–Schmidt procedure. The modified Gram–Schmidt procedure is known to yield nearly orthogonal vectors if the vectors being orthogonalized are not too nearly linearly dependent. In the special case where the vectors are almost linearly dependent, the modified Gram–Schmidt procedure can be replaced by Householder transformations, at the cost of some extra arithmetic [139]. In this case, one would expect the basis vectors generated in the GMRES method to be almost orthogonal and the approximate solution obtained to be nearly optimal, at least in the space spanned by these vectors. For discussions of the effect of rounding errors on the GMRES method, see [33, 70, 2].

This is not the case for the CG and MINRES algorithms, which use short recurrences to generate orthogonal basis vectors for the Krylov subspace. The proof of orthogonality, and hence of the optimality of the approximate solution, relies on induction (e.g., Theorems 2.3.1 and 2.3.2 and the arguments after the Lanczos algorithm in section 2.5), and such arguments may be destroyed by the effects of finite precision arithmetic. In fact, the basis vectors generated by the Lanczos algorithm (or the residual vectors generated by the CG algorithm) in finite precision arithmetic frequently lose orthogonality completely and may even become linearly dependent! In such cases, the approximate solutions generated by the CG and MINRES algorithms are *not* the optimal approximations from the Krylov subspace, and it is not clear that any of the results from Chapter 3 should hold.

In this chapter we show why the nonorthogonal vectors generated by the Lanczos algorithm can still be used effectively for solving linear systems and which of the results from Chapter 3 can and cannot be expected to hold (to a close approximation) in finite precision arithmetic. It is shown that for both the MINRES and CG algorithms, the 2-norm of the residual is essentially determined by the *tridiagonal matrix* produced in the finite precision Lanczos computation. This tridiagonal matrix is, of course, quite different from the one that would be produced in exact arithmetic. It follows, however, that if the same tridiagonal matrix would be produced by the exact Lanczos algorithm applied to some other problem, then exact arithmetic bounds on the residual for that problem will hold for the finite precision computation. In order to establish exact arithmetic bounds for the different problem, it is necessary to have some information about the eigenvalues of the new coefficient matrix. Here we make use of results already established in the literature about the eigenvalues of the new coefficient matrix, but we do not include the proofs.

The analysis presented here is by no means a complete rounding-error analysis of the algorithms given in Chapter 2. As anyone who has done a rounding-error analysis knows, the arguments can quickly become complicated and tedious. Here we attempt to present some of the more interesting aspects of the error analysis for the CG and MINRES algorithms, without becoming bogged down in the details. We consider a hypothetical implementation of these algorithms for which the analysis is easier and refer to the literature for arguments about the precise nature of the roundoff terms at each step.

This analysis deals with the rate at which the A-norm of the error in the CG algorithm and the 2-norm of the residual in the MINRES algorithm are reduced before the ultimately attainable accuracy is achieved. A separate issue is the level of accuracy that can be attained if the iteration is carried out for sufficiently many steps. This question is discussed in section 7.3.

4.1. Some Numerical Examples.

To illustrate the numerical behavior of the CG algorithm of section 2.3, we have applied this algorithm to linear systems $Ax = b$ with coefficient matrices of the form $A = U\Lambda U^H$, where U is a random orthogonal matrix and $\Lambda = \text{diag}(\lambda_1, \ldots, \lambda_n)$, where

$$(4.1) \qquad \lambda_i = \lambda_1 + \frac{i-1}{n-1}(\lambda_n - \lambda_1)\rho^{n-i}, \quad i = 2, \ldots, n-1,$$

and the parameter ρ is chosen between 0 and 1. For $\rho = 1$, the eigenvalues are uniformly spaced, and for smaller values of ρ the eigenvalues are tightly clustered at the lower end of the spectrum and are far apart at the upper end. We set $n = 24$, $\lambda_1 = .001$, and $\lambda_n = 1$. A random right-hand side and zero initial guess were used in all cases. Figure 4.1a shows a plot of the A-norm of the error versus the iteration number for $\rho = .4, .6, .8, 1$. Experiments were performed using double precision Institute of Electrical and Electronics Engineers (IEEE) arithmetic, with machine precision $\epsilon \approx 1.1e{-}16$. Figure 4.1b shows what these curves would look like if exact arithmetic had been used. (Exact arithmetic can be simulated in the CG algorithm by saving all of the basis vectors in the Lanczos algorithm and explicitly orthogonalizing against them at every step. This is how the data for Figure 4.1b was produced.)

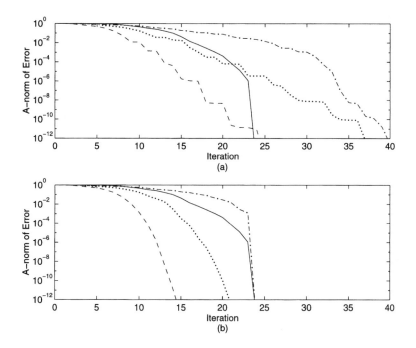

FIG. 4.1. *CG convergence curves for* (a) *finite precision arithmetic and* (b) *exact arithmetic.* $\rho = 1$ *solid,* $\rho = .8$ *dash–dot,* $\rho = .6$ *dotted,* $\rho = .4$ *dashed.*

Note that although the theory of Chapter 3 guarantees that the exact solution is obtained after $n = 24$ steps, the computations with $\rho = .6$ and $\rho = .8$ do not generate good approximate solutions by step 24. It is at about step 31 that the error in the $\rho = .8$ computation begins to decrease rapidly. The $\rho = .4$ computation has reduced the A-norm of the error to $1.e-12$ by step 24, but the corresponding exact arithmetic calculation would have reduced it to this level after just 14 steps. In contrast, for $\rho = 1$ (the case with equally spaced eigenvalues), the exact and finite precision computations behave very similarly. In all cases, the finite precision computation eventually finds a good approximate solution, but it is clear that estimates of the number of iterations required to do so cannot be based on the error bounds of Chapter 3. In this chapter we develop error bounds that hold in finite precision arithmetic.

4.2. The Lanczos Algorithm.

When the Lanczos algorithm is implemented in finite precision arithmetic, the recurrence of section 2.5 is perturbed slightly. It is replaced by a recurrence that can be written in matrix form as

$$(4.2) \qquad AQ_k = Q_k T_k + \beta_k q_{k+1} \xi_k^T + F_k = Q_{k+1} T_{k+1,k} + F_k,$$

where the columns of F_k represent the rounding errors at each step. Let ϵ denote the machine precision and define

$$(4.3) \qquad \epsilon_0 \equiv 2(n+4)\epsilon, \quad \epsilon_1 \equiv 2(7 + m \parallel |A| \parallel / \|A\|) \, \epsilon,$$

where m is the maximum number of nonzeros in any row of A. Under the assumptions that

$$(4.4) \qquad \epsilon_0 < \frac{1}{12}, \quad k(3\epsilon_0 + \epsilon_1) < 1,$$

and ignoring higher order terms in ϵ, Paige [109] showed that the rounding error matrix F_k satisfies

$$(4.5) \qquad \|F_k\| \leq \sqrt{k} \, \epsilon_1 \, \|A\|.$$

Paige also showed that the coefficient formulas in the Lanczos algorithm can be implemented sufficiently accurately to ensure that

$$(4.6) \qquad |q_j^T q_j - 1| \leq 2\epsilon_0,$$

$$(4.7) \qquad \beta_j \leq \|A\| \, (1 + (2n + 6)\epsilon + j(3\epsilon_0 + \epsilon_1)).$$

We will assume throughout that the inequalities (4.4) and hence (4.5–4.7) hold.

Although the individual roundoff terms are tiny, their effect on the recurrence (4.2) may be great. The Lanczos vectors may lose orthogonality and even become linearly dependent. The recurrence coefficients generated in finite precision arithmetic may be quite different from those that would be generated in exact arithmetic.

4.3. A Hypothetical MINRES/CG Implementation.

Although the computed Lanczos vectors may not be orthogonal, one might still consider using them in the CG or MINRES algorithms for solving linear systems; that is, one could still choose an approximate solution x_k of the form

$$(4.8) \qquad x_k = x_0 + Q_k y_k,$$

where y_k solves the least squares problem

$$(4.9) \qquad \min_y \|\beta \xi_1 - T_{k+1,k} y\|, \quad \beta \equiv \|r_0\|$$

for the MINRES method or the linear system

$$(4.10) \qquad T_k y = \beta \xi_1$$

for the CG algorithm. Of course, in practice, one does not first compute the Lanczos vectors and then apply formulas (4.8–4.10), since this would require saving all of the Lanczos vectors. Still, it is reasonable to try and separate the effects of roundoff on the three-term Lanczos recurrence from that on other aspects of the (implicit) evaluation of (4.8–4.10). It is the effect of using the nonorthogonal vectors produced by a finite precision Lanczos computation that is analyzed here, so from here on we assume that formulas (4.8–4.10) hold exactly, where Q_k, T_k, and $T_{k+1,k}$ satisfy (4.2).

The residual in the CG algorithm, which we denote here as r_k^C, then satisfies

$$
\begin{aligned}
r_k^C &= r_0 - AQ_k y_k^C = r_0 - (Q_k T_k + \beta_k q_{k+1} \xi_k^T + F_k) y_k^C \\
&= -\beta_k q_{k+1} \xi_k^T y_k^C - F_k y_k^C \\
&= -\beta \, (\beta_k q_{k+1} \xi_k^T T_k^{-1} \xi_1 + F_k y_k^C / \beta),
\end{aligned}
$$

where y_k^C denotes the solution to (4.10). The 2-norm of the residual satisfies

$$
\|q_{k+1}\| \, |\beta_k \xi_k^T T_k^{-1} \xi_1| - \|F_k\| \, \|y_k^C\|/\beta \le \|r_k^C\|/\|r_0\|
$$

(4.11) $$\le \|q_{k+1}\| \, |\beta_k \xi_k^T T_k^{-1} \xi_1| + \|F_k\| \, \|y_k^C\|/\beta.$$

Using (4.5) and (4.6), this becomes

$$
\sqrt{1 + 2\epsilon_0} \, |\beta_k \xi_k^T T_k^{-1} \xi_1| - \sqrt{k} \, \epsilon_1 \, \|A\| \, \|y_k^C\|/\beta \le \|r_k^C\|/\|r_0\|
$$

(4.12) $$\le \sqrt{1 + 2\epsilon_0} \, |\beta_k \xi_k^T T_k^{-1} \xi_1| + \sqrt{k} \, \epsilon_1 \, \|A\| \, \|y_k^C\|/\beta.$$

It follows that at steps k, where $\sqrt{k} \, \epsilon_1 \, \|A\| \, \|y_k^C\|/\beta$ is much smaller than the residual norm, the 2-norm of the residual is essentially determined by the tridiagonal matrix T_k and the next recurrence coefficient β_k.

The residual in the MINRES algorithm, which we denote here as r_k^M, satisfies

$$
\begin{aligned}
r_k^M &= r_0 - AQ_k y_k^M = r_0 - (Q_{k+1} T_{k+1,k} + F_k) y_k^M \\
&= Q_{k+1}(\beta \xi_1 - T_{k+1,k} y_k^M) - F_k y_k^M,
\end{aligned}
$$

where y_k^M denotes the solution to (4.9). The 2-norm of the residual satisfies

(4.13) $$\|r_k^M\|/\|r_0\| \le \|Q_{k+1}\| \, \|\xi_1 - T_{k+1,k} y_k^M / \beta\| + \|F_k\| \, \|y_k^M\|/\beta.$$

It follows from (4.6) that

$$
\|Q_{k+1}\| \le \sqrt{(1 + 2\epsilon_0)(k + 1)},
$$

so, with (4.5), we have

$$
\|r_k^M\|/\|r_0\| \le \sqrt{(1 + 2\epsilon_0)(k + 1)} \, \|\xi_1 - T_{k+1,k} y_k^M / \beta\|
$$

(4.14) $$+\sqrt{k} \, \epsilon_1 \, \|A\| \, \|y_k^M\|/\beta.$$

It follows that at steps k, where $\sqrt{k} \, \epsilon_1 \, \|A\| \, \|y_k^M\|/\beta$ is tiny compared to the residual norm, the 2-norm of the residual is essentially bounded, to within a possible factor of $\sqrt{k+1}$ (which is usually an overestimate), by an expression involving only the $k + 1$-by-k tridiagonal matrix $T_{k+1,k}$.

Thus, for both the MINRES and CG algorithms, the 2-norm of the residual (or at least a realistic bound on the 2-norm of the residual) is essentially determined by the recurrence coefficients computed in the finite precision

Lanczos computation and stored in the tridiagonal matrix $T_{k+1,k}$. Suppose the exact Lanczos algorithm, applied to a matrix (or linear operator) $\mathcal{A}$ with initial vector φ_1, generates the same tridiagonal matrix $T_{k+1,k}$. It would follow that the 2-norm of the residual r_k^M or r_k^C in the finite precision computation would be approximately the same as the 2-norm of the residual ν_k^M or ν_k^C in the exact MINRES or CG algorithm for solving the linear system (or operator equation) $\mathcal{A}\chi = \varphi$, with right-hand side $\varphi = \beta\varphi_1$; in this case, we would have

$$\|\nu_k^C\|/\|r_0\| = |\beta_k \xi_k^T T_k^{-1} \xi_1|, \quad \|\nu_k^M\|/\|r_0\| = \|\xi_1 - T_{k+1,k} y_k^M/\beta\|.$$

Compare with (4.12) and (4.14).

Note also that if T is any Hermitian tridiagonal matrix (even an infinite one) whose upper left $k + 1$-by-k block is $T_{k+1,k}$, then the exact Lanczos algorithm applied to T with initial vector ξ_1 will generate the matrix $T_{k+1,k}$ at step k. This follows because the reduction of a Hermitian matrix to tridiagonal form (with nonnegative off-diagonal entries) is uniquely determined once the initial vector is set.

With this observation, we can now use results about the convergence of the exact MINRES and CG algorithms applied to any such matrix T to derive bounds on the residuals r_k^M and r_k^C in the finite precision computation. To do this, one must have some information about the eigenvalues of such a matrix T.

4.4. A Matrix Completion Problem.

With the arguments of the previous section, the problem of bounding the residual norm in finite precision CG and MINRES computations becomes a matrix completion problem: given the $k + 1$-by-k tridiagonal matrix $T_{k+1,k}$ generated by a finite precision Lanczos computation, find a Hermitian tridiagonal matrix T with $T_{k+1,k}$ as its upper left block,

$$T = \begin{pmatrix} \alpha_1 & \beta_1 & & & & & & \\ \beta_1 & \ddots & \ddots & & & & & \\ & \ddots & \ddots & \beta_{k-1} & & & & \\ & & \beta_{k-1} & \alpha_k & \beta_k & & & \\ & & & \beta_k & * & * & & \\ & & & & * & \ddots & \ddots & \\ & & & & & \ddots & \ddots & * \\ & & & & & & * & * \end{pmatrix},$$

whose eigenvalues are related to those of A in such a way that the exact arithmetic error bounds of Chapter 3 yield useful results about the convergence of the exact CG or MINRES algorithms applied to linear systems with coefficient matrix T. In this section we state such results but refer the reader to the literature for their proofs.

4.4.1. Paige's Theorem. The following result of Paige [110] shows that the eigenvalues of T_{k+1}, the tridiagonal matrix generated at step $k+1$ of a finite precision Lanczos computation, lie essentially between the largest and smallest eigenvalues of A.

THEOREM 4.4.1 (Paige). *The eigenvalues $\theta_i^{(j)}$, $i = 1, \ldots, j$ of the tridiagonal matrix T_j satisfy*

$$(4.15) \quad \lambda_1 - j^{5/2}\epsilon_2\|A\| \leq \theta_i^{(j)} \leq \lambda_n + j^{5/2}\epsilon_2\|A\|, \quad \epsilon_2 = \sqrt{2}\,\max\{6\epsilon_0, \epsilon_1\},$$

where λ_1 is the smallest eigenvalue of A, λ_n is the largest eigenvalue of A, and ϵ_0 and ϵ_1 are defined in (4.3).

Using this result with the arguments of the previous section, we obtain the following result for Hermitian positive definite matrices A.

THEOREM 4.4.2. *Let A be a Hermitian positive definite matrix with eigenvalues $\lambda_1 \leq \cdots \leq \lambda_n$ and assume that $\lambda_1 - (k+1)^{5/2}\epsilon_2\|A\| > 0$. Let r_k^M and r_k^C denote the residuals at step k of the MINRES and CG computations satisfying (4.8) and (4.9) or (4.10), respectively, where Q_k, T_k, and $T_{k+1,k}$ satisfy (4.2). Then*

$$(4.16) \qquad \|r_k^C\|/\|r_0\| \leq \sqrt{1 + 2\epsilon_0}\,2\sqrt{\tilde{\kappa}}\left(\frac{\sqrt{\tilde{\kappa}} - 1}{\sqrt{\tilde{\kappa}} + 1}\right)^k + \sqrt{k}\,\tilde{\kappa}\,\epsilon_1,$$

$$(4.17) \quad \|r_k^M\|/\|r_0\| \leq \sqrt{(1 + 2\epsilon_0)(k+1)}\,2\left(\frac{\sqrt{\tilde{\kappa}} - 1}{\sqrt{\tilde{\kappa}} + 1}\right)^k + \sqrt{k}\,\tilde{\kappa}\,\epsilon_1,$$

where

$$(4.18) \qquad\qquad \tilde{\kappa} = \frac{\lambda_n + (k+1)^{5/2}\epsilon_2\|A\|}{\lambda_1 - (k+1)^{5/2}\epsilon_2\|A\|}$$

and ϵ_0, ϵ_1, and ϵ_2 are defined in (4.3) and (4.15).

Proof. It follows from Theorem 4.4.1 that T_k is nonsingular, and since $y_k^C = T_k^{-1}\beta\xi_1$, we have for the second term on the right-hand side of (4.12)

$$\|A\|\,\|y_k^C\|/\beta \leq \|A\|\,\|T_k^{-1}\| \leq \frac{\lambda_n}{\lambda_1 - k^{5/2}\epsilon_2\|A\|} \leq \tilde{\kappa}.$$

Since the expression $|\beta_k\xi_k^T T_k^{-1}\xi_1|$ in (4.12) is the size of the residual at step k of the exact CG algorithm applied to a linear system with coefficient matrix T_{k+1} and right-hand side ξ_1 and since the eigenvalues of T_{k+1} satisfy (4.15), it follows from (3.8) that

$$|\beta_k\xi_k^T T_k^{-1}\xi_1| \leq 2\sqrt{\tilde{\kappa}}\left(\frac{\sqrt{\tilde{\kappa}} - 1}{\sqrt{\tilde{\kappa}} + 1}\right)^k,$$

where $\tilde{\kappa}$ is given by (4.18). Here we have used the fact that since the expression in (3.8) bounds the reduction in the T_{k+1}-norm of the error in the exact

CG iterate, the reduction in the 2-norm of the residual for this exact CG iterate is bounded by $\sqrt{\tilde{\kappa}}$ times the expression in (3.8); i.e., $\|Bv\|/\|Bw\| \leq \sqrt{\kappa(B)} \, \|v\|_B/\|w\|_B$ for any vectors v and w and any positive definite matrix B. Making these substitutions into (4.12) gives the desired result (4.16).

For the MINRES algorithm, it can be seen from the Cauchy interlace theorem (Theorem 1.3.12) applied to $T_{k+1,k}T_{k+1,k}^H$ that the smallest singular value of $T_{k+1,k}$ is greater than or equal to the smallest eigenvalue of T_k. Consequently, we have

$$\|y_k^M\| \leq \beta \, \|T_k^{-1}\|,$$

so, similar to the CG algorithm, the second term on the right-hand side of (4.14) satisfies

$$\|A\| \, \|y_k^M\|/\beta \leq \|A\| \, \|T_k^{-1}\| \leq \frac{\lambda_n}{\lambda_1 - k^{5/2}\epsilon_2\|A\|} \leq \tilde{\kappa}.$$

Since the expression $\|\xi_1 - T_{k+1,k}y_k^M/\beta\|$ in (4.14) is the size of the residual at step k of the exact MINRES algorithm applied to the linear system $T_{k+1}\chi = \xi_1$, where the eigenvalues of T_{k+1} satisfy (4.15), it follows from (3.12) that

$$\|\xi_1 - T_{k+1,k}y_k^M/\beta\| \leq 2 \left(\frac{\sqrt{\tilde{\kappa}} - 1}{\sqrt{\tilde{\kappa}} + 1}\right)^k.$$

Making these substitutions into (4.14) gives the desired result (4.17). $\square$

Theorem 4.4.2 shows that, at least to a close approximation, the exact arithmetic residual bounds based on the size of the Chebyshev polynomial on the *interval* from the smallest to the largest eigenvalue of A hold in finite precision arithmetic as well. Exact arithmetic bounds such as (3.11) and (3.12) for $\ell > 0$, based on approximation on *discrete* subsets of the eigenvalues of A may fail, however, as may the sharp bounds (3.6) and (3.7). This was illustrated in section 4.1. Still, stronger bounds than (4.16) and (4.17) may hold in finite precision arithmetic, and such bounds are derived in the next subsection.

4.4.2. A Different Matrix Completion.
Paige's theorem about the eigenvalues of T_{k+1} lying essentially between the smallest and largest eigenvalues of A is of little use in the case of indefinite A, since in that case, T_{k+1} could be singular. Moreover, we would like to find a completion T of $T_{k+1,k}$ whose eigenvalues can be more closely related to the discrete eigenvalues of A in order to obtain finite precision analogues of the sharp error bounds (3.6) and (3.7).

It was shown by Greenbaum that $T_{k+1,k}$ can be extended to a larger Hermitian tridiagonal matrix T whose eigenvalues all lie in tiny intervals about the eigenvalues of A [65], the size of the intervals being a function of the machine precision. Unfortunately, the proven bound on the interval size appears to be a large overestimate. The bound on the interval size established in [65] involves a number of constants as well as a factor of the form $n^3 k^2 \sqrt{\epsilon} \|A\|$ or, in some cases, $n^3 k \epsilon^{1/4} \|A\|$, but better bounds are believed possible.

Suppose the eigenvalues of such a matrix T have been shown to lie in intervals of width δ about the eigenvalues of A. One can then relate the size of the residual at step k of a finite precision computation to the maximum value of the minimax polynomial on the union of tiny intervals containing the eigenvalues of T, using the same types of arguments as given in Theorem 4.4.2.

THEOREM 4.4.3. *Let A be a Hermitian matrix with eigenvalues $\lambda_1 \leq \cdots \leq \lambda_n$ and let $T_{k+1,k}$ be the $k + 1$-by-k tridiagonal matrix generated by a finite precision Lanczos computation. Assume that there exists a Hermitian tridiagonal matrix T, with $T_{k+1,k}$ as its upper left $k + 1$-by-k block, whose eigenvalues all lie in the intervals*

$$(4.19) \qquad S = \bigcup_{i=1}^{n} [\lambda_i - \delta, \lambda_i + \delta],$$

where none of the intervals contains the origin. Let d denote the distance from the origin to the set S. Then the MINRES residual r_k^M satisfies

$$\|r_k^M\|/\|r_0\| \leq \sqrt{(1 + 2\epsilon_0)(k + 1)} \min_{p_k} \max_{z \in S} |p_k(z)|$$

$$(4.20) \qquad\qquad + 2\sqrt{k} \, (\lambda_n/d) \, \epsilon_1.$$

If A is positive definite, then the CG residual r_k^C satisfies

$$\|r_k^C\|/\|r_0\| \leq \sqrt{1 + 2\epsilon_0} \sqrt{(\lambda_n + \delta)/d} \min_{p_k} \max_{z \in S} |p_k(z)|$$

$$(4.21) \qquad\qquad + \sqrt{k} \, (\lambda_n/d) \, \epsilon_1.$$

Proof. Since the expression $\|\xi_1 - T_{k+1,k} y_k^M / \beta\|$ in (4.14) is the size of the residual at step k of the exact MINRES algorithm applied to the linear system $T\chi = \xi_1$, where the eigenvalues of T lie in S, it follows from (3.7) that

$$\|\xi_1 - T_{k+1,k} y_k^M / \beta\| \leq \min_{p_k} \max_{z \in S} |p_k(z)|.$$

To bound the second term in (4.14), note that the approximate solution generated at step k of this corresponding exact MINRES calculation is of the form $\chi_k = \mathcal{Q}_k y_k^M / \beta$, where the columns of $\mathcal{Q}_k$ are orthonormal and the vector y_k^M is the same one generated by the finite precision computation. It follows that $\|y_k^M\|/\beta = \|\chi_k\|$. Since the 2-norm of the residual decreases monotonically in the exact algorithm, we have

$$\|\xi_1 - T\chi_k\| \leq 1 \implies \|T^{-1}\xi_1 - \chi_k\| \leq \|T^{-1}\| \implies \|\chi_k\| \leq 2\|T^{-1}\|.$$

Making these substitutions in (4.14) gives

$$\|r_k^M\|/\|r_0\| \leq \sqrt{(1 + 2\epsilon_0)(k + 1)} \min_{p_k} \max_{z \in S} |p_k(z)| + 2\sqrt{k} \, \epsilon_1 \, \|A\| \, \|T^{-1}\|,$$

from which the desired result (4.20) follows.

When A, and hence T, is positive definite, the expression $|\beta_k \xi_k^T T_k^{-1} \xi_1|$ in (4.12) is the size of the residual at step k of the exact CG algorithm applied to the linear system $T\chi = \xi_1$. It follows from (3.6) that

$$|\beta_k \xi_k^T T_k^{-1} \xi_1| \leq \sqrt{\kappa(T)} \min_{p_k} \max_{z \in S} |p_k(z)|,$$

where the factor $\sqrt{\kappa(T)} = \sqrt{(\lambda_n + \delta)/d}$ must be included, since this gives a bound on the 2-norm of the residual instead of the T-norm of the error. The second term in (4.12) can be bounded as in Theorem 4.4.2. Since $y_k^C = T_k^{-1} \beta \xi_1$ and since, by the Cauchy interlace theorem, the smallest eigenvalue of T_k is greater than or equal to that of T, we have

$$\|A\| \, \|y_k^C\|/\beta \leq \|A\| \, \|T_k^{-1}\| \leq \lambda_n/d.$$

Making these substitutions in (4.12) gives the desired result (4.21).   □

Theorem 4.4.3 shows that, to a close approximation, the exact arithmetic residual bounds based on the size of the minimax polynomial on the *discrete* set of eigenvalues of A can be replaced, in finite precision arithmetic, by the size of the minimax polynomial on the union of *tiny intervals* in (4.19). Bounds such as (3.11) and (3.12) for $\ell > 0$ will not hold in finite precision arithmetic. Instead, if A has a few large outlying eigenvalues, one must consider a polynomial that is the product of one with enough roots in the outlying intervals to ensure that it is tiny throughout these intervals, with a lower-degree Chebyshev polynomial on the remainder of the spectrum. The maximum value of this polynomial throughout the set S in (4.19) provides an error bound that holds in finite precision arithmetic. It is still advantageous, in finite precision arithmetic, to have most eigenvalues concentrated in a small interval with just a few outliers (as opposed to having eigenvalues everywhere throughout the larger interval), but the advantages are less than in exact arithmetic.

It is shown in [65] that not only the residual norm bound (4.21) but also the corresponding bound on the A-norm of the error,

$$(4.22) \qquad \|e_k^C\|_A/\|e_0\|_A \leq \min_{p_k} \max_{z \in S} |p_k(z)|,$$

holds to a close approximation in finite precision arithmetic. In Figure 4.2, this error bound is plotted along with the actual A-norm of the error in a finite precision computation with a random right-hand side and zero initial guess for the case $\rho = .6$ described in section 4.1. The interval width δ was taken to be $1.e - 15$, or about 10ϵ. For comparison, the sharp error bound for exact arithmetic (3.6) is also shown in Figure 4.2. It is evident that the bound (4.22) is applicable to the finite precision computation and that it gives a reasonable estimate of the actual error when the initial residual is random.

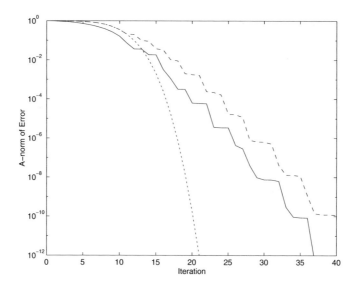

FIG. 4.2. *Exact arithmetic error bound (dotted), finite precision arithmetic error bound (assuming $\delta = 1.e - 15$) (dashed), and actual error in a finite precision computation (solid).*

4.5. Orthogonal Polynomials.

The theorems of the previous section identified the behavior of the first k steps of the CG and MINRES algorithms in finite precision arithmetic with that of the first k steps of the exact algorithms applied to a different problem. (That is, the tridiagonal matrices generated during the first k steps are the same, and so the residual norms agree to within the factors given in the theorems.) Of course, if the bound δ on the interval size in Theorem 4.4.3 were *independent* of k, then this would imply that the identity (between tridiagonal matrices generated in finite precision arithmetic and those generated by the exact algorithm applied to a linear operator with eigenvalues contained in intervals of width δ about the eigenvalues of A) would hold for arbitrarily many steps. It is not known whether the assumption of Theorem 4.4.3 can be satisfied for some small value of δ that does not depend on k.

The analysis of section 4.4 is somewhat unusual in linear algebra. Normally, the approximate solution generated by a finite precision computation is identified with the exact solution of a nearby problem of the *same dimension*. The matrix T in Theorem 4.4.3 can be of any dimension greater than or equal to $k+1$. It represents a *nearby* problem only in the sense that its eigenvalues lie close to those of A. The arguments of the previous sections have a somewhat more natural interpretation in terms of orthogonal polynomials.

The 3-term recurrence of the Lanczos algorithm (in exact arithmetic) implicitly constructs the orthonormal polynomials for a certain set of weights on the eigenvalues of the matrix—the weights being the squared components of the initial vector in the direction of each eigenvector of A. To see this, let $A = U\Lambda U^H$ be an eigendecomposition of A and let $\hat{q}_j = U^H q_j$, where the vectors q_j, $j = 1, 2, \ldots$, are the Lanczos vectors generated by the algorithm in section 2.5. Then, following the algorithm of section 2.5, we have

(4.23) $$\hat{q}_{j+1} = \beta_j^{-1}(\Lambda\hat{q}_j - \alpha_j\hat{q}_j - \beta_{j-1}\hat{q}_{j-1}),$$

where

$$\alpha_j = \langle\Lambda\hat{q}_j - \beta_{j-1}\hat{q}_{j-1}, \hat{q}_j\rangle, \quad \beta_j = \|\Lambda\hat{q}_j - \alpha_j\hat{q}_j - \beta_{j-1}\hat{q}_{j-1}\|.$$

It follows that the ith component of $\hat{q}_{j+1}$ is equal to a certain jth-degree polynomial, say, $\psi_j(z)$, evaluated at λ_i, times the ith component of $\hat{q}_1$. The polynomials $\psi_j(z)$, $j = 1, 2, \ldots$, satisfy

(4.24) $$\psi_j(z) = \beta_j^{-1}(z\psi_{j-1}(z) - \alpha_j\psi_{j-1}(z) - \beta_{j-1}\psi_{j-2}(z)),$$

where $\psi_{-1}(z) \equiv 0$, $\psi_0(z) \equiv 1$. If we define the w-inner product of two polynomials ϕ and ψ by

(4.25) $$\langle\phi(z), \psi(z)\rangle_w \equiv \sum_{i=1}^{n} \phi(\lambda_i)\psi(\lambda_i)\hat{q}_{i1}^2,$$

where $\hat{q}_{i1}$ is the ith component of $\hat{q}_1$, then the coefficients in the Lanczos algorithm are given by

(4.26) $$\alpha_j = \langle z\psi_{j-1}(z) - \beta_{j-1}\psi_{j-2}(z), \psi_{j-1}(z)\rangle_w,$$

(4.27) $$\beta_j = \|z\psi_{j-1}(z) - \alpha_j\psi_{j-1}(z) - \beta_{j-1}\psi_{j-2}(z)\|_w,$$

where $\|\phi(z)\|_w \equiv \langle\phi(z), \phi(z)\rangle_w^{1/2}$.

Equation (4.24) with coefficient formulas (4.26–4.27) defines the orthonormal polynomials for the measure corresponding to the w-inner product in (4.25). It follows from the orthonormality of the Lanczos vectors that these polynomials satisfy $\langle\psi_j(z), \psi_k(z)\rangle_w = \delta_{jk}$.

A perturbation vector f_j in the Lanczos algorithm, due to finite precision arithmetic, corresponds to a perturbation $\hat{f}_j = U^H f_j$ of the same size in (4.23). The finite precision analogue of recurrence (4.24) is

(4.28) $$\psi_j(z) = \beta_j^{-1}(z\psi_{j-1}(z) - \alpha_j\psi_{j-1}(z) - \beta_{j-1}\psi_{j-2}(z) - \zeta_j(z)),$$

where $\zeta_j(\lambda_i)\hat{q}_{i1} = \hat{f}_{ij}$. If we imagine that the coefficient formulas (4.26–4.27) hold *exactly* in finite precision arithmetic, where the functions $\psi_j(z)$ now come from the perturbed recurrence (4.28), we still find that the intended orthogonality relation $\langle\psi_j(z), \psi_k(z)\rangle_w = \delta_{jk}$ may fail completely. (It is reasonable to assume that the coefficient formulas (4.26–4.27) hold exactly, since they can be implemented very accurately and any differences between the exact formulas and the computed values can be included in the perturbation term $\zeta_j(z)$.)

It is possible that some coefficient β_j in a finite precision Lanczos computation will be exactly 0 and that the recurrence will terminate, but this is unlikely. If β_j is not 0, then it is positive because of formula (4.27). It follows from a theorem due to Favard [48] that the recurrence coefficients constructed in a finite precision Lanczos computation are the exact recurrence coefficients for the orthonormal polynomials corresponding to *some* nonnegative measure. That is, if we define $\rho_{-1}(z) \equiv 0$, $\rho_0(z) \equiv 1$, and

$$(4.29) \qquad \rho_j(z) = \beta_j^{-1}(z\rho_{j-1}(z) - \alpha_j\rho_{j-1}(z) - \beta_{j-1}\rho_{j-2}(z))$$

for $j = 1, 2, \ldots$, where α_j and β_j are defined by (4.26–4.28), then we have the following theorem.

THEOREM 4.5.1 (Favard). *If the coefficients β_j in (4.29) are all positive and the α_j's are real, then there is a measure $d\omega(z)$ such that*

$$\int \rho_j(z)\rho_k(z)\,d\omega(z) = \delta_{jk}$$

for all $j, k = 0, 1, \ldots, \infty$.

The measure $d\omega(z)$ in Favard's theorem is (substantially) uniquely determined, whereas there are infinitely many measures for which the first k polynomials $\rho_0, \ldots, \rho_{k-1}$ are orthonormal. One such measure—a measure with weights on the eigenvalues of T_{k+1}, the weights being the squared first components of each eigenvector of T_{k+1}—was given in section 4.4.1, and another such measure—a measure with weights on points in tiny intervals about the eigenvalues of A—was given in section 4.4.2. It was also shown in [65] that the weight on each interval is approximately equal to the original weight on the corresponding eigenvalue of A; that is, the squared component of $\hat{q}_1$. Thus, the matrix completion result of section 4.4.2 can also be stated in the following way: when one attempts to construct the first k orthonormal polynomials for a measure corresponding to weights on discrete points using the Lanczos algorithm, what one actually obtains are the first k orthonormal polynomials for a slightly different measure—one in which the weights are smeared out over tiny intervals about the original points. Exactly how the weights are distributed over these intervals depends on exactly what rounding errors occur (not just on their size).

It remains an open question whether the measure defined by Favard's theorem has its support in such tiny intervals (i.e., whether δ in Theorem 4.4.3 can be taken to be small and independent of k). If this is not the case, it might still be possible to show that the measure in Favard's theorem is *tiny* everywhere outside such intervals.

Comments and Additional References.

It should come as no surprise that the Lanczos vectors and tridiagonal matrix can be used for many purposes besides solving linear systems. For example, the eigenvalues of T_k can be taken as approximations to some of the eigenvalues of A. It is given as an exercise to show that the orthogonal polynomials defined in section 4.5 are the characteristic polynomials of the successive tridiagonal matrices generated by the Lanczos algorithm. This interpretation enables one to use known properties of the roots of orthogonal polynomials to describe the eigenvalue approximations. In finite precision arithmetic, the fact that the polynomials (or at least a finite sequence of these polynomials) are orthogonal with respect to a slightly smeared-out version of the original measure helps to explain the nature of eigenvalue approximations generated during a finite precision Lanczos computation. Depending on how the tiny intervals of Theorem 4.4.3 are distributed, the corresponding orthogonal polynomials might have several roots in some of the intervals before having any roots in some of the others. This is usually the case with an interval corresponding to a large well-separated eigenvalue. This explains the observed phenomenon of multiple close approximations to some eigenvalues appearing in finite precision Lanczos computations before any approximations to some of the other eigenvalues appear.

The Lanczos vectors and tridiagonal matrix can also be used very effectively to compute the matrix exponential $\exp(tA)\varphi$, which is the solution at time t to the system of differential equations $y' = Ay$, $y(0) = \varphi$. Similar arguments to those used here show why the nonorthogonal vectors generated by a finite precision Lanczos computation can still be used effectively for this purpose [34]. For a number of other applications, including discussions of the effects of finite precision arithmetic, see, for example, [35, 60].

The effect of rounding errors on the CG algorithm has been a subject of concern since the algorithm was first introduced in 1952 by Hestenes and Stiefel [79]. It was recognized at that time that the algorithm did not always behave the way exact arithmetic theory predicted. For example, Engeli et al. [43] applied the CG method (without a preconditioner) to the biharmonic equation and observed that convergence did not occur until well after step n. For this and other reasons, the algorithm did not gain widespread popularity at that time.

With the idea of preconditioning in the CG method, interest in this algorithm was revived in the early 1970's [115, 27], and it quickly became the method of choice for computations involving large Hermitian positive definite matrices. Whatever the effect of roundoff, it was observed that the method performed very well in comparison to other iterative methods. Further attempts were made to explain the success of the method, mostly using the interpretation given in section 2.3 that the algorithm minimizes the A-norm of the error in a plane that includes the direction of steepest descent. Using this argument, Wozniakowski [143] showed that a special version of the CG algorithm does, indeed, reduce the A-norm of the error at each step by at least as much as a steepest descent step, even in finite precision arithmetic. Cullum and Willoughby [30] proved a similar result for a more standard version of the algorithm. Still, a more global approach was needed to explain why the CG algorithm converges so much faster than the method of steepest descent; e.g., it converges at least as fast as the Chebyshev algorithm. Paige's work on the Lanczos algorithm [109] provided a key in this direction. A number of analyses were developed to explain the behavior of the CG algorithm using information from the entire computation (i.e., the matrix equation (2.23)), instead of just one or two steps (e.g., [35, 62, 65, 121]). The analogy developed in this chapter, identifying the finite precision computation with the exact algorithm applied to a different matrix, appears to be very effective in explaining and predicting the behavior of the CG algorithm in finite precision arithmetic [71]. The numerical examples presented in section 4.1 were first presented in [126].

Exercises.

4.1. Show that the orthonormal polynomials defined by (4.24) are the *characteristic polynomials* of the tridiagonal matrices generated by the Lanczos algorithm.

4.2. How must the error bound you derived in Exercise 3.1 for a matrix with a small, well-separated eigenvalue be modified for finite precision arithmetic? Does the finite precision error bound differ more from that of exact arithmetic in the case when a positive definite coefficient matrix has one eigenvalue much smaller than the others or in the case when it has one eigenvalue much larger than the others? (This comparison can be used to explain why one preconditioner might be considered better based on exact arithmetic theory, but a different preconditioner might perform better in actual computations. See [133] for a comparison of incomplete Cholesky and modified incomplete Cholesky decompositions, which will be discussed in Chapter 11.)

Chapter 5

BiCG and Related Methods

Since the GMRES method for non-Hermitian problems requires increasing amounts of work and storage per iteration, it is important to consider other methods with a fixed amount of work and storage, even though they will require more iterations to reduce the 2-norm of the residual to a given level. Several such methods have already been presented, e.g., simple iteration, Orthomin(j), and GMRES(j). All have the possibility of failure: simple iteration may diverge, Orthomin(j) may encounter an undefined coefficient, and both Orthomin(j) and GMRES(j) may stagnate (cease to reduce the residual norm).

In this chapter we consider several other iteration methods that, in practice, have often been found to perform better than the previously listed algorithms. These algorithms also have the possibility of failure, although that can be alleviated through the use of *look-ahead*. With look-ahead, however, the methods no longer require a fixed amount of work and storage per iteration. The work and storage grows with the number of look-ahead steps, just as it grows in GMRES. Unfortunately, there are no a priori theoretical estimates comparing the error at each step of these methods to that of the optimal GMRES approximation, unless an unlimited number of look-ahead steps are allowed. This problem is discussed further in Chapter 6.

5.1. The Two-Sided Lanczos Algorithm.

When the matrix A is Hermitian, the Gram–Schmidt procedure for constructing an orthonormal basis for the Krylov space of A reduces to a 3-term recurrence. Unfortunately, this is not the case when A is non-Hermitian. One can, however, use a *pair* of 3-term recurrences, one involving A and the other involving A^H, to construct *biorthogonal* bases for the Krylov spaces corresponding to A and A^H. Let $\mathcal{K}_k(B, v)$ denote the Krylov space span$\{v, Bv, \ldots, B^{k-1}v\}$. Then one constructs two sets of vectors— $v_1, \ldots, v_k \in \mathcal{K}_k(A, r_0)$ and $w_1, \ldots, w_k \in \mathcal{K}_k(A^H, \hat{r}_0)$—such that $\langle v_i, w_j \rangle = 0$ for $i \neq j$. This procedure is called the *two-sided Lanczos algorithm*.

> **Two-Sided Lanczos Algorithm** (without look-ahead).
>
> Given r_0 and $\hat{r}_0$ with $\langle r_0, \hat{r}_0 \rangle \neq 0$, set $v_1 = r_0/\|r_0\|$ and $w_1 = \hat{r}_0/\langle \hat{r}_0, v_1 \rangle$. Set $\beta_0 = \gamma_0 = 0$ and $v_0 = w_0 = 0$. For $j = 1, 2, \ldots$,
>
> Compute Av_j and $A^H w_j$.
>
> Set $\alpha_j = \langle Av_j, w_j \rangle$.
>
> Compute $\tilde{v}_{j+1} = Av_j - \alpha_j v_j - \beta_{j-1} v_{j-1}$ and $\tilde{w}_{j+1} = A^H w_j - \bar{\alpha}_j w_j - \bar{\gamma}_{j-1} w_{j-1}$.
>
> Set $\gamma_j = \|\tilde{v}_{j+1}\|$ and $v_{j+1} = \tilde{v}_{j+1}/\gamma_j$.
>
> Set $\beta_j = \langle v_{j+1}, \tilde{w}_{j+1} \rangle$ and $w_{j+1} = \tilde{w}_{j+1}/\bar{\beta}_j$.

Here we have given the non-Hermitian Lanczos formulation that scales so that each basis vector v_j has norm 1 and $\langle w_j, v_j \rangle = 1$. The scaling of the basis vectors can be chosen differently. Another formulation of the algorithm uses the ordinary transpose A^T, instead of A^H.

Letting V_k be the matrix with columns $v_1, \ldots, v_k$ and W_k be the matrix with columns $w_1, \ldots, w_k$, this pair of recurrences can be written in matrix form as

$$(5.1) \qquad AV_k = V_k T_k + \gamma_k v_{k+1} \xi_k^T = V_{k+1} T_{k+1,k},$$

$$(5.2) \qquad A^H W_k = W_k T_k^H + \bar{\beta}_k w_{k+1} \xi_k^T = W_{k+1} \hat{T}_{k+1,k},$$

where T_k is the k-by-k tridiagonal matrix of recurrence coefficients

$$T_k = \begin{pmatrix} \alpha_1 & \beta_1 & & & \\ \gamma_1 & \ddots & \ddots & & \\ & \ddots & \ddots & \beta_{k-1} \\ & & \gamma_{k-1} & \alpha_k \end{pmatrix}.$$

The $k+1$-by-k matrices $T_{k+1,k}$ and $\hat{T}_{k+1,k}$ have T_k and T_k^H, respectively, as their top k-by-k blocks, and their last rows consist of zeros except for the last entry, which is γ_k and $\bar{\beta}_k$, respectively. The biorthogonality condition implies that

$$(5.3) \qquad V_k^H W_k = I.$$

Note that if $A = A^H$ and $\hat{r}_0 = r_0$, then the two-sided Lanczos recurrence reduces to the ordinary Hermitian Lanczos process.

THEOREM 5.1.1. *If the two-sided Lanczos vectors are defined at steps* $1, \ldots, k+1$, *that is, if* $\langle v_j, w_j \rangle \neq 0$, $j = 1, \ldots, k+1$, *then*

$$(5.4) \qquad \langle v_i, w_j \rangle = 0 \quad \forall i \neq j, \quad i, j \leq k+1.$$

Proof. Assume that (5.4) holds for $i, j \leq k$. The choice of the coefficients β_j and γ_j assures that for all j, $\langle w_j, v_j \rangle = 1$ and $\|v_j\| = 1$. By construction of the coefficient α_k, we have, using the induction hypothesis,

$$\langle \tilde{v}_{k+1}, w_k \rangle = \langle Av_k, w_k \rangle - \alpha_k = 0,$$

$$\langle \tilde{w}_{k+1}, v_k \rangle = \langle A^H w_k, v_k \rangle - \bar{\alpha}_k = 0.$$

Using the recurrences for $\tilde{v}_{k+1}$ and $\tilde{w}_k$ along with the induction hypothesis, we have

$$\langle \tilde{v}_{k+1}, w_{k-1} \rangle = \langle Av_k, w_{k-1} \rangle - \beta_{k-1} = \langle v_k, A^H w_{k-1} \rangle - \beta_{k-1} = \langle v_k, \tilde{w}_k \rangle - \beta_{k-1} = 0,$$

and, similarly, it follows that $\langle \tilde{w}_{k+1}, v_{k-1} \rangle = 0$. Finally, for $j < k - 1$, we have

$$\langle \tilde{v}_{k+1}, w_j \rangle = \langle Av_k, w_j \rangle = \langle v_k, A^H w_j \rangle = \langle v_k, \tilde{w}_{j+1} \rangle = 0,$$

and, similarly, it is seen that $\langle \tilde{w}_{k+1}, v_j \rangle = 0$. Since v_{k+1} and w_{k+1} are just multiples of $\tilde{v}_{k+1}$ and $\tilde{w}_{k+1}$, the result (5.4) is proved. $\quad\square$

The vectors generated by the two-sided Lanczos process can become undefined in two different situations. First, if $\tilde{v}_{j+1} = 0$ or $\tilde{w}_{j+1} = 0$, then the Lanczos algorithm has found an invariant subspace. If $\tilde{v}_{j+1} = 0$, then the right Lanczos vectors $v_1, \ldots, v_j$ form an A-invariant subspace. If $\tilde{w}_{j+1} = 0$, then the left Lanczos vectors $w_1, \ldots, w_j$ form an A^H-invariant subspace. This is referred to as *regular termination.*

The second case, referred to as *serious breakdown*, occurs when $\langle \tilde{v}_{j+1}, \tilde{w}_{j+1} \rangle = 0$ but neither $\tilde{v}_{j+1} = 0$ nor $\tilde{w}_{j+1} = 0$. In this case, nonzero vectors $v_{j+1} \in \mathcal{K}_{j+1}(A, r_0)$ and $w_{j+1} \in \mathcal{K}_{j+1}(A^H, \hat{r}_0)$ satisfying $\langle v_{j+1}, w_i \rangle = \langle w_{j+1}, v_i \rangle = 0$ for all $i \leq j$ simply do not exist. Note, however, that while such vectors may not exist at step $j + 1$, at some later step $j + \ell$ there may be nonzero vectors $v_{j+\ell} \in \mathcal{K}_{j+\ell}(A, r_0)$ and $w_{j+\ell} \in \mathcal{K}_{j+\ell}(A^H, \hat{r}_0)$ such that $v_{j+\ell}$ is orthogonal to $\mathcal{K}_{j+\ell-1}(A^H, \hat{r}_0)$ and $w_{j+\ell}$ is orthogonal to $\mathcal{K}_{j+\ell-1}(A, r_0)$. Procedures that simply skip steps at which the Lanczos vectors are undefined and construct the Lanczos vectors for the steps at which they are defined are referred to as *look-ahead* Lanczos methods. We will not discuss look-ahead Lanczos methods here but refer the reader to [101, 113, 20] for details.

5.2. The Biconjugate Gradient Algorithm.

Let us assume for the moment that the Lanczos recurrence does not break down. (From here on, we will refer to the two-sided Lanczos algorithm as simply the Lanczos algorithm, since it is the only Lanczos algorithm for non-Hermitian matrices.) Then the basis vectors might be used to approximate the solution of a linear system, as was done in the Hermitian case. If x_k is taken to be of the form

(5.5) $$x_k = x_0 + V_k y_k,$$

then there are several natural ways to choose the vector y_k. One choice is to force $r_k = r_0 - AV_k y_k$ to be orthogonal to $w_1, \ldots, w_k$. This leads to the equation

$$W_k^H r_k = W_k^H r_0 - W_k^H A V_k y_k = 0.$$

It follows from (5.1) and the biorthogonality condition (5.3) that $W_k^H A V_k = T_k$ and that $W_k^H r_0 = \beta \xi_1$, $\beta = \|r_0\|$, so the equation for y_k becomes

$$(5.6) \qquad\qquad T_k y_k = \beta \xi_1.$$

When A is Hermitian and $\hat{r}_0 = r_0$, this reduces to the CG algorithm, as was described in section 2.5. If T_k is singular, this equation may have no solution. In this case, there is no approximation x_k of the form (5.5) for which $W_k^H r_k = 0$. Note that this type of failure is *different* from the possible breakdown of the underlying Lanczos recurrence. The Lanczos vectors may be well defined, but if the tridiagonal matrix is singular or near singular, an algorithm that attempts to solve this linear system will have difficulty. An algorithm that attempts to generate approximations of the form (5.5), where y_k satisfies (5.6), is called the *biconjugate gradient* (BiCG) algorithm.

The BiCG algorithm can be derived from the non-Hermitian Lanczos process and the *LDU*-factorization of the tridiagonal matrix T_k in much the same way that the CG algorithm was derived from the Hermitian Lanczos process in section 2.5. As for CG, failure of the BiCG method occurs when a singular tridiagonal matrix is encountered, and, with the standard implementation of the algorithm, one cannot recover from a singular tridiagonal matrix at one step, even if later tridiagonal matrices are well conditioned. We will not carry out this derivation since it is essentially the same as that in section 2.5 but will simply state the algorithm as follows.

Biconjugate Gradient Algorithm (BiCG).

Given x_0, compute $r_0 = b - Ax_0$, and set $p_0 = r_0$. Choose $\hat{r}_0$ such that $\langle r_0, \hat{r}_0 \rangle \neq 0$, and set $\hat{p}_0 = \hat{r}_0$. For $k = 1, 2, \ldots$

Set $x_k = x_{k-1} + a_{k-1} p_{k-1}$, where $a_{k-1} = \frac{\langle r_{k-1}, \hat{r}_{k-1} \rangle}{\langle A p_{k-1}, \hat{p}_{k-1} \rangle}$.

Compute $r_k = r_{k-1} - a_{k-1} A p_{k-1}$ and $\hat{r}_k = \hat{r}_{k-1} - \bar{a}_{k-1} A^H \hat{p}_{k-1}$.

Set $p_k = r_k + b_{k-1} p_{k-1}$ and $\hat{p}_k = \hat{r}_k + \bar{b}_{k-1} \hat{p}_{k-1}$, where $b_{k-1} = \frac{\langle r_k, \hat{r}_k \rangle}{\langle r_{k-1}, \hat{r}_{k-1} \rangle}$.

A better implementation can be derived from that of the QMR algorithm described in section 5.3.

5.3. The Quasi-Minimal Residual Algorithm.

In the *quasi-minimal residual* (QMR) algorithm, the approximate solution x_k is again taken to be of the form (5.5), but now y_k is chosen to minimize a quantity that is closely related to the 2-norm of the residual. Since $r_k = r_0 - AV_ky_k$, we can write

$$(5.7) \qquad r_k = V_{k+1}(\beta\xi_1 - T_{k+1,k}y_k),$$

so the norm of r_k satisfies

$$(5.8) \qquad \|r_k\| \le \|V_{k+1}\| \cdot \|\beta\xi_1 - T_{k+1,k}y_k\|.$$

Since the columns of V_{k+1} are not orthogonal, it would be difficult to choose y_k to minimize $\|r_k\|$, but y_k can easily be chosen to minimize the second factor in (5.8). Since the columns of V_{k+1} each have norm one, the first factor in (5.8) satisfies $\|V_{k+1}\| \le \sqrt{k+1}$. In the QMR method, y_k solves the least squares problem

$$(5.9) \qquad \min_y \|\beta\xi_1 - T_{k+1,k}y\|,$$

which always has a solution, even if the tridiagonal matrix T_k is singular. Thus the QMR iterates are defined provided that the underlying Lanczos recurrence does not break down.

The norm of the QMR residual can be related to that of the optimal GMRES residual as follows.

THEOREM 5.3.1 (Nachtigal [101]). *If r_k^G denotes the GMRES residual at step k and r_k^Q denotes the QMR residual at step k, then*

$$(5.10) \qquad \|r_k^Q\| \le \kappa(V_{k+1})\,\|r_k^G\|,$$

where V_{k+1} is the matrix of basis vectors for the space $\mathcal{K}_{k+1}(A, r_0)$ constructed by the Lanczos algorithm and $\kappa(\cdot)$ denotes the condition number.

Proof. The GMRES residual is also of the form (5.7), but the vector y_k^G is chosen to minimize the 2-norm of the GMRES residual. It follows that

$$\|r_k^G\| \ge \sigma_{min}(V_{k+1})\,\min_y \|\beta\xi_1 - T_{k+1,k}y\|,$$

where $\sigma_{min}(V_{k+1})$ is the smallest singular value. Combining this with inequality (5.8) for the QMR residual gives the desired result (5.10). $\square$

Unfortunately, the condition number of the basis vectors V_{k+1} produced by the non-Hermitian Lanczos algorithm cannot be bounded a priori. This matrix may be ill conditioned, even if the Lanczos vectors are well defined. If one could devise a short recurrence that would generate *well-conditioned* basis vectors, then one could use the quasi-minimization strategy (5.9) to solve the problem addressed in Chapter 6.

The actual implementation of the QMR algorithm, without saving all of the Lanczos vectors, is similar to that of the MINRES algorithm described in section 2.5. The least squares problem (5.9) is solved by factoring the $k+1$-by-k matrix $T_{k+1,k}$ into the product of a $k+1$-by-$k+1$ unitary matrix F^H

and a $k + 1$-by-k upper triangular matrix R. This is accomplished by using k Givens rotations $F_1, \ldots, F_k$, where F_i rotates the unit vectors ξ_i and ξ_{i+1} through angle θ_i. Since $T_{k+1,k}$ is tridiagonal, R has the form

$$
R = \begin{pmatrix}
\rho_1 & \sigma_1 & \tau_1 & & & \\
& \ddots & \ddots & \ddots & & \\
& & \ddots & \ddots & \tau_{k-2} & \\
& & & \ddots & \sigma_{k-1} & \\
& & & & \rho_k & \\
0 & \cdots & \cdots & \cdots & 0 &
\end{pmatrix}.
$$

The QR decomposition of $T_{k+1,k}$ is easily updated from that of $T_{k,k-1}$. To obtain R, first premultiply the last column of $T_{k+1,k}$ by the rotations from steps $k - 2$ and $k - 1$ to obtain a matrix of the form

$$
F_{k-1}F_{k-2}(F_{k-3} \cdots F_1)T_{k+1,k} = \begin{pmatrix}
x & x & x & & 0 \\
& \ddots & \ddots & \ddots & \\
& & \ddots & \ddots & x \\
& & & \ddots & x \\
& & & & d \\
0 & \cdots & \cdots & \cdots & h
\end{pmatrix},
$$

where the x's denote nonzeros and where the $(k + 1, k)$-entry, h, is just γ_k, since this entry is unaffected by the previous rotations. The next rotation, F_k, is chosen to annihilate this entry by setting $c_k = |d|/\sqrt{|d|^2 + |h|^2}$; $\bar{s}_k = c_k h/d$ if $d \neq 0$, and $c_k = 0$, $\bar{s}_k = 1$ if $d = 0$. To solve the least squares problem, the successive rotations are also applied to the right-hand side vector $\beta \xi_1$ to obtain $g = F_k \cdots F_1 \beta \xi_1$. Clearly, g differs from the corresponding vector at step $k - 1$ only in positions k and $k + 1$. If $R_{k \times k}$ denotes the top k-by-k block of R and $g_{k \times 1}$ denotes the first k entries of g, then the solution to the least squares problem is the solution of the triangular linear system

$$
R_{k \times k} y_k = g_{k \times 1}.
$$

In order to update the iterates x_k, we define auxiliary vectors

$$
P_k \equiv (p_0, \ldots, p_{k-1}) \equiv V_k R_{k \times k}^{-1}.
$$

Then since

$$
x_k = x_0 + V_k y_k = x_0 + P_k g_{k \times 1}
$$

and

$$
x_{k-1} = x_0 + P_{k-1} g_{k-1 \times 1},
$$

we can write

$$(5.11) \qquad x_k = x_{k-1} + a_{k-1}p_{k-1},$$

where a_{k-1} is the kth entry of g. Finally, from the equation $P_k R_{k \times k} = V_k$, we can update the auxiliary vectors using

$$(5.12) \qquad p_{k-1} = \frac{1}{\rho_k}(v_k - \sigma_{k-1}p_{k-2} - \tau_{k-2}p_{k-3}).$$

This leads to the following implementation of the QMR algorithm.

Algorithm 5. Quasi-Minimal Residual Method (QMR) (without look-ahead).

Given x_0, compute $r_0 = b - Ax_0$ and set $v_1 = r_0/\|r_0\|$.
Given $\hat{r}_0$, set $w_1 = \hat{r}_0/\|\hat{r}_0\|$. Initialize $\xi = (1, 0, \ldots, 0)^T$, $\beta = \|r_0\|$.
For $k = 1, 2, \ldots$,

Compute v_{k+1}, w_{k+1}, $\alpha_k \equiv T(k,k)$, $\beta_k \equiv T(k, k+1)$, and $\gamma_k \equiv T(k+1, k)$, using the two-sided Lanczos algorithm.

Apply F_{k-2} and F_{k-1} to the last column of T; that is,

$$\begin{pmatrix} T(k-2, k) \\ T(k-1, k) \end{pmatrix} \leftarrow \begin{pmatrix} c_{k-2} & s_{k-2} \\ -\bar{s}_{k-2} & c_{k-2} \end{pmatrix} \begin{pmatrix} 0 \\ T(k-1, k) \end{pmatrix}, \quad \text{if } k > 2,$$

$$\begin{pmatrix} T(k-1, k) \\ T(k, k) \end{pmatrix} \leftarrow \begin{pmatrix} c_{k-1} & s_{k-1} \\ -\bar{s}_{k-1} & c_{k-1} \end{pmatrix} \begin{pmatrix} T(k-1, k) \\ T(k, k) \end{pmatrix}, \quad \text{if } k > 1.$$

Compute the kth rotation c_k and s_k, to annihilate the $(k+1, k)$ entry of T.[1]

Apply kth rotation to ξ and to last column of T:

$$\begin{pmatrix} \xi(k) \\ \xi(k+1) \end{pmatrix} \leftarrow \begin{pmatrix} c_k & s_k \\ -\bar{s}_k & c_k \end{pmatrix} \begin{pmatrix} \xi(k) \\ 0 \end{pmatrix}$$

$$T(k, k) \leftarrow c_k T(k, k) + s_k T(k+1, k), \quad T(k+1, k) \leftarrow 0.$$

Compute $p_{k-1} = [v_k - T(k-1, k)p_{k-2} - T(k-2, k)p_{k-3}]/T(k, k)$, where undefined terms are zero for $k \leq 2$.

Set $x_k = x_{k-1} + \beta\xi(k)p_{k-1}$.

---

[1]The formula is $c_k = |T(k, k)|/\sqrt{|T(k, k)|^2 + |T(k+1, k)|^2}$, $\bar{s}_k = c_k T(k+1, k)/T(k, k)$, but a more robust implementation should be used. See, for example, BLAS routine DROTG [32].

5.4. Relation Between BiCG and QMR.

The observant reader may have noted that the solutions of the linear system (5.6) and the least squares problem (5.9) are closely related. Hence one might expect a close relationship between the residual norms in the BiCG and QMR algorithms. Here we establish such a relationship, assuming that the Lanczos vectors are well defined and that the tridiagonal matrix in (5.6) is nonsingular.

We begin with a general theorem about the relationship between upper Hessenberg linear systems and least squares problems. Let H_k, $k = 1, 2, \ldots$, denote a family of upper Hessenberg matrices, where H_k is k-by-k and H_{k-1} is the $k-1$-by-$k-1$ principal submatrix of H_k. For each k, define the $k+1$-by-k matrix $H_{k+1,k}$ by

$$H_{k+1,k} = \begin{pmatrix} H_k \\ h_{k+1,k}\xi_k^T \end{pmatrix}$$

The matrix $H_{k+1,k}$ can be factored in the form $F^H R$, where F is a $k+1$-by-$k+1$ unitary matrix and R is a $k+1$ by k upper triangular matrix. This factorization can be performed using plane rotations in the manner described for the GMRES algorithm in section 2.4:

$$(F_k \cdots F_1) H_{k+1,k} = R, \quad \text{where } F_i = \begin{pmatrix} I_{i-1} & & & \\ & c_i & -s_i & \\ & s_i & c_i & \\ & & & I_{k-i} \end{pmatrix}$$

Note that the first $k-1$ sines and cosines s_i, c_i, $i = 1, \ldots, k-1$, are those used in the factorization of $H_{k,k-1}$.

Let $\beta > 0$ be given and assume that H_k is nonsingular. Let $\tilde{y}_k$ denote the solution of the linear system $H_k y = \beta \xi_1$, and let y_k denote the solution of the least squares problem $\min_y \|H_{k+1,k} y - \beta \xi_1\|$. Finally, let

$$\tilde{\nu}_k = H_{k+1,k}\tilde{y}_k - \beta\xi_1, \quad \nu_k = H_{k+1,k}y_k - \beta\xi_1.$$

LEMMA 5.4.1. *Using the above notation, the norms of ν_k and $\tilde{\nu}_k$ are related to the sines and cosines of the Givens rotations by*

(5.13) $$\|\nu_k\| = \beta|s_1 s_2 \cdots s_k| \quad \text{and} \quad \|\tilde{\nu}_k\| = \beta\frac{1}{|c_k|}|s_1 s_2 \cdots s_k|.$$

It follows that

(5.14) $$\|\tilde{\nu}_k\| = \frac{\|\nu_k\|}{\sqrt{1 - (\|\nu_k\|/\|\nu_{k-1}\|)^2}}.$$

Proof. The least squares problem with the extended Hessenberg matrix $H_{k+1,k}$ can be written in the form

$$\min_y \|H_{k+1,k}y - \beta\xi_1\| = \min_y \|F(H_{k+1,k}y - \beta\xi_1)\| = \min_y \|Ry - \beta F\xi_1\|,$$

and the solution y_k is determined by solving the upper triangular linear system with coefficient matrix equal to the top k-by-k block of R and right-hand side equal to the first k entries of $\beta F \xi_1$. The remainder $R y_k - \beta F \xi_1$ is therefore zero except for the last entry, which is just the last entry of $-\beta F \xi_1 = -\beta (F_k \cdots F_1) \xi_1$, which is easily seen to be $-\beta s_1 \cdots s_k$. This establishes the first equality in (5.13).

For the linear system solution $\tilde{y}_k = H_k^{-1} \beta \xi_1$, we have

$$\tilde{\nu}_k = H_{k+1,k} H_k^{-1} \beta \xi_1 - \beta \xi_1,$$

which is zero except for the last entry, which is $\beta h_{k+1,k}$ times the $(k,1)$-entry of H_k^{-1}. Now H_k can be factored in the form $\tilde{F}^H \tilde{R}$, where $\tilde{F} = \tilde{F}_{k-1} \cdots \tilde{F}_1$ and $\tilde{F}_i$ is the k-by-k principal submatrix of F_i. The matrix $H_{k+1,k}$, after applying the first $k - 1$ plane rotations, has the form

$$
(F_{k-1} \cdots F_1) \, H_{k+1,k} =
\begin{pmatrix}
x & x & \cdots & x \\
 & x & \cdots & x \\
 & & \ddots & \vdots \\
 & & & r \\
 & & & h
\end{pmatrix}
$$

where r is the (k,k)-entry of $\tilde{R}$ and $h = h_{k+1,k}$. The kth rotation is chosen to annihilate the nonzero entry in the last row:

$$c_k = r / \sqrt{r^2 + h^2}, \qquad s_k = -h / \sqrt{r^2 + h^2}.$$

Note that r and c_k are nonzero since H_k is nonsingular.

We can write $H_k^{-1} = \tilde{R}^{-1} \tilde{F}$, and the $(k,1)$-entry of this is $1/r$ times the $(k,1)$-entry of $\tilde{F} = \tilde{F}_{k-1} \cdots \tilde{F}_1$, and this is just $s_1 \cdots s_{k-1}$. It follows that the nonzero entry of $\tilde{\nu}_k$ is $\beta(h_{k+1,k}/r) s_1 \cdots s_{k-1}$. Finally, using the fact that $|s_k/c_k| = |h/r| = |h_{k+1,k}/r|$, we obtain the second equality in (5.13).

From (5.13) it is clear that

$$\frac{\|\nu_k\|}{\|\nu_{k-1}\|} = |s_k|, \qquad \frac{\|\tilde{\nu}_k\|}{\|\nu_k\|} = 1/|c_k|.$$

The result (5.14) follows upon replacing $|c_k|$ by $\sqrt{1 - |s_k|^2}$. □

An immediate consequence of this lemma is the following relationship between the BiCG residual r_k^B and the quantity

$$(5.15) \qquad z_k^Q \equiv \beta \xi_1 - T_{k+1,k} y_k^Q,$$

which is related to the residual r_k^Q in the QMR algorithm:

$$(5.16) \qquad r_k^Q = V_{k+1} z_k^Q, \qquad \|r_k^Q\| \leq \sqrt{k+1} \|z_k^Q\|.$$

We will refer to z_k^Q as the QMR *quasi-residual*—the vector whose norm is actually minimized in the QMR algorithm.

TABLE 5.1

Relation between QMR quasi-residual norm reduction and ratio of BiCG residual norm to QMR quasi-residual norm.

$\|z_k^Q\|/\|z_{k-1}^Q\|$	$\|r_k^B\|/\|z_k^Q\|$
.5	1.2
.9	2.3
.99	7.1
.9999	70.7
.999999	707

THEOREM 5.4.1. *Assume that the Lanczos vectors at steps 1 through k are defined and that the tridiagonal matrix generated by the Lanczos algorithm at step k is nonsingular. Then the BiCG residual r_k^B and the QMR quasi-residual z_k^Q are related by*

$$(5.17) \qquad \|r_k^B\| = \frac{\|z_k^Q\|}{\sqrt{1 - (\|z_k^Q\|/\|z_{k-1}^Q\|)^2}}.$$

Proof. From (5.1), (5.5), and (5.6), it follows that the BiCG residual can be written in the form

$$\begin{aligned}
r_k^B &= r_0 - AV_k y_k^B \\
&= r_0 - V_{k+1} T_{k+1,k} y_k^B \\
&= V_{k+1}(\beta \xi_1 - T_{k+1,k} T_k^{-1} \beta \xi_1).
\end{aligned}$$

The quantity in parentheses has only one nonzero entry (in its $(k+1)$st position), and since $\|v_{k+1}\| = 1$, we have

$$(5.18) \qquad \|r_k^B\| = \|\beta \xi_1 - T_{k+1,k} T_k^{-1} \beta \xi_1\|.$$

The desired result now follows from Lemma 5.4.1 and the definition (5.15) of z_k^Q. □

In most cases, the quasi-residual norms and the actual residual norms in the QMR algorithm are of the same order of magnitude. Inequality (5.16) shows that the latter can exceed the former by at most a factor of $\sqrt{k+1}$, and a bound in the other direction is given by

$$\|r_k^Q\| \geq \sigma_{min}(V_{k+1})\|z_k^Q\|,$$

where σ_{min} denotes the smallest singular value. While it is possible that $\sigma_{min}(V_{k+1})$ is very small (especially in finite precision arithmetic), it is unlikely that $\|r_k^Q\|$ would be much smaller than $\|z_k^Q\|$. The vector y_k^Q is chosen to satisfy the least squares problem (5.9), without regard to the matrix V_{k+1}.

Theorem 5.4.1 shows that if the QMR quasi-residual norm is reduced by a significant factor at step k, then the BiCG residual norm will be approximately

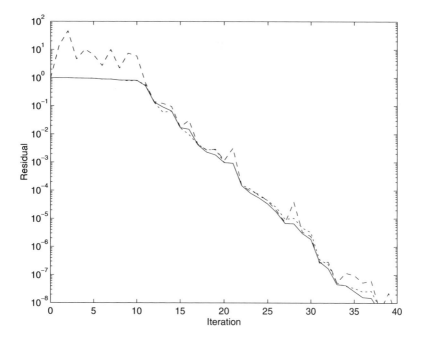

FIG. 5.1. *BiCG residual norms (dashed), QMR residual norms (dotted), and QMR quasi-residual norms (solid).*

equal to the QMR quasi-residual norm at step k, since the denominator in the right-hand side of (5.17) will be close to 1. If the QMR quasi-residual norm remains almost constant, however, then the denominator in the right-hand side of (5.17) will be close to 0, and the BiCG residual norm will be much larger. Table 5.1 shows the relation between the QMR quasi-residual norm reduction and the ratio of BiCG residual norm to QMR quasi-residual norm. Note that the QMR quasi-residual norm must be *very* flat before the BiCG residual norm is orders-of-magnitude larger.

Figure 5.1 shows a plot of the logarithms of the norms of the BiCG residuals (dashed line), the QMR residuals (dotted line), and the QMR quasi-residuals (solid line) versus iteration number for a simple example problem. The matrix A, a real 103-by-103 matrix, was taken to have 50 pairs of complex conjugate eigenvalues, randomly distributed in the rectangle $[1, 2] \times [-i, i]$, and 3 additional real eigenvalues at 4, .5, and -1. A random matrix V was generated and A was set equal to VDV^{-1}, where D is a block-diagonal matrix with 3 1-by-1 blocks corresponding to the separated real eigenvalues of A and 50 2-by-2 blocks of the form

$$\begin{pmatrix} a & b \\ -b & a \end{pmatrix}$$

corresponding to the pairs of eigenvalues $a \pm ib$.

In this example, the QMR residual and quasi-residual norm curves are barely distinguishable. As predicted by Theorem 5.4.1, peaks in the BiCG residual norm curve correspond to plateaus in the QMR convergence curve. At steps where the QMR quasi-residual norm is reduced by a large factor, the BiCG residual norm is reduced by an even greater amount so that it "catches up" with QMR.

Relation (5.17) implies roughly that the BiCG and QMR algorithms will either both converge well or both perform poorly for a given problem. While the QMR quasi-residual norm cannot increase during the iteration, it is no more useful to have a near constant residual norm than it is to have an increasing one. The analysis here assumes exact arithmetic, however. In finite precision arithmetic, one might expect that a very large intermediate iterate (corresponding to a very large residual norm) could lead to inaccuracy in the final approximation, and, indeed, such a result was established in [66]. This will be discussed further in section 7.3. A thorough study of the effect of rounding errors on the BiCG and QMR algorithms has not been carried out, however. Since the BiCG and QMR algorithms require essentially the same amount of work and storage per iteration and since the QMR quasi-residual norm is always less than or equal to the BiCG residual norm, it seems reasonable to choose QMR over BiCG, although the difference may not be great.

5.5.   The Conjugate Gradient Squared Algorithm.

The BiCG and QMR algorithms require multiplication by both A and A^H at each step. This means extra work, and, additionally, it is sometimes much less convenient to multiply by A^H than it is to multiply by A. For example, there may be a special formula for the product of A with a given vector when A represents, say, a Jacobian, but a corresponding formula for the product of A^H with a given vector may not be available. In other cases, data may be stored on a parallel machine in such a way that multiplication by A is efficient but multiplication by A^H involves extra communication between processors. For these reasons it is desirable to have an iterative method that requires multiplication only by A and that generates good approximate solutions from the Krylov spaces of dimension equal to the number of matrix–vector multiplications. A method that attempts to do this is the conjugate gradient squared (CGS) method.

Returning to the BiCG algorithm of section 5.2, note that we can write

$$r_k = \varphi_k(A)r_0, \quad \hat{r}_k = \bar{\varphi}_k(A^H)\hat{r}_0,$$

$$p_k = \psi_k(A)r_0, \quad \hat{p}_k = \bar{\psi}_k(A^H)\hat{r}_0$$

for certain kth-degree polynomials φ_k and ψ_k. If the algorithm is converging well, then $\|\varphi_k(A)r_0\|$ is small and one might expect that $\|\varphi_k^2(A)r_0\|$ would be even smaller. If $\varphi_k^2(A)r_0$ could be computed with about the same amount of work as $\varphi_k(A)r_0$, then this would likely result in a faster converging algorithm. This is the idea of CGS.

Rewriting the BiCG recurrence in terms of these polynomials, we see that

(5.19) $$\varphi_k(A)r_0 = \varphi_{k-1}(A)r_0 - a_{k-1}A\psi_{k-1}(A)r_0,$$

(5.20) $$\psi_k(A)r_0 = \varphi_k(A)r_0 + b_k\psi_{k-1}(A)r_0,$$

where

(5.21) $$a_{k-1} = \frac{\langle \varphi_{k-1}(A)r_0, \bar{\varphi}_{k-1}(A^H)\hat{r}_0 \rangle}{\langle A\psi_{k-1}(A)r_0, \bar{\psi}_{k-1}(A^H)\hat{r}_0 \rangle} = \frac{\langle \varphi_{k-1}^2(A)r_0, \hat{r}^0 \rangle}{\langle A\psi_{k-1}^2(A)r_0, \hat{r}_0 \rangle},$$

(5.22) $$b_k = \frac{\langle \varphi_k(A)r_0, \bar{\varphi}_k(A^H)\hat{r}_0 \rangle}{\langle \varphi_{k-1}(A)r_0, \bar{\varphi}_{k-1}(A^H)\hat{r}_0 \rangle} = \frac{\langle \varphi_k^2(A)r_0, \hat{r}_0 \rangle}{\langle \varphi_{k-1}^2(A)r_0, \hat{r}_0 \rangle}.$$

Note that the coefficients can be computed if we know $\hat{r}_0$ and $\varphi_j^2(A)r_0$ and $\psi_j^2(A)r_0$, $j = 1, 2, \ldots$.

From (5.19–5.20), it can be seen that the polynomials $\varphi_k(z)$ and $\psi_k(z)$ satisfy the recurrences

$$\varphi_k(z) = \varphi_{k-1}(z) - a_{k-1}z\psi_{k-1}(z), \quad \psi_k(z) = \varphi_k(z) + b_k\psi_{k-1}(z),$$

and squaring both sides gives

$$\varphi_k^2(z) = \varphi_{k-1}^2(z) - 2a_{k-1}z\varphi_{k-1}(z)\psi_{k-1}(z) + a_{k-1}^2z^2\psi_{k-1}^2(z),$$

$$\psi_k^2(z) = \varphi_k^2(z) + 2b_k\varphi_k(z)\psi_{k-1}(z) + b_k^2\psi_{k-1}^2(z).$$

Multiplying φ_k by the recurrence for ψ_k gives

$$\varphi_k(z)\psi_k(z) = \varphi_k^2(z) + b_k\varphi_k(z)\psi_{k-1}(z),$$

and multiplying the recurrence for φ_k by ψ_{k-1} gives

$$\begin{aligned} \varphi_k(z)\psi_{k-1}(z) &= \varphi_{k-1}(z)\psi_{k-1}(z) - a_{k-1}z\psi_{k-1}^2(z) \\ &= \varphi_{k-1}^2(z) + b_{k-1}\varphi_{k-1}(z)\psi_{k-2}(z) - a_{k-1}z\psi_{k-1}^2(z). \end{aligned}$$

Defining

$$\Phi_k \equiv \varphi_k^2, \quad \Theta_k \equiv \varphi_k\psi_{k-1}, \quad \Psi_{k-1} \equiv \psi_{k-1}^2,$$

these recurrences become

$$\begin{aligned} \Phi_k(z) &= \Phi_{k-1}(z) - 2a_{k-1}z(\Phi_{k-1}(z) + b_{k-1}\Theta_{k-1}(z)) + a_{k-1}^2z^2\Psi_{k-1}(z), \\ \Theta_k(z) &= \Phi_{k-1}(z) + b_{k-1}\Theta_{k-1}(z) - a_{k-1}z\Psi_{k-1}(z), \\ \Psi_k(z) &= \Phi_k(z) + 2b_k\Theta_k(z) + b_k^2\Psi_{k-1}. \end{aligned}$$

Let $r_k^S = \Phi_k(A)r_0$, $p_k^S = \Psi_k(A)r_0$, and $q_k^S = \Theta_k(A)r_0$. Then the following algorithm generates an approximate solution x_k^S with the required residual r_k^S.

Conjugate Gradient Squared Algorithm (CGS).

Given $x_0 \equiv x_0^S$, compute $r_0 \equiv r_0^S = b - Ax_0^S$, set $u_0^S = r_0^S$, $p_0^S = r_0^S$, $q_0^S = 0$, and $v_0^S = Ap_0^S$. Set an arbitrary vector $\hat{r}_0$. For $k = 1, 2, \ldots$

Compute $q_k^S = u_{k-1}^S - a_{k-1}v_{k-1}^S$, where $a_{k-1} = \dfrac{\langle r_{k-1}^S, \hat{r}_0 \rangle}{\langle v_{k-1}^S, \hat{r}_0 \rangle}$.

Set $x_k^S = x_{k-1}^S + a_{k-1}(u_{k-1}^S + q_k^S)$.

Then $r_k^S = r_{k-1}^S - a_{k-1}A(u_{k-1}^S + q_k^S)$.

Compute $u_k^S = r_k^S + b_k q_k^S$, where $b_k = \frac{\langle r_k^S, \hat{r}_0 \rangle}{\langle r_{k-1}^S, \hat{r}_0 \rangle}$.

Set $p_k^S = u_k^S + b_k(q_k^S + b_k p_{k-1}^S)$ and $v_k^S = A p_k^S$.

The CGS method requires two matrix–vector multiplications at each step but no multiplications by the Hermitian transpose. For problems where the BiCG method converges well, CGS typically requires only about half as many steps and, therefore, half the work of BiCG (assuming that multiplication by A or A^H requires the same amount of work). When the norm of the BiCG residual increases at a step, however, that of the CGS residual usually increases by approximately the square of the increase of the BiCG residual norm. The CGS convergence curve may therefore show wild oscillations that can sometimes lead to numerical instabilities.

5.6. The BiCGSTAB Algorithm.

To avoid the large oscillations in the CGS convergence curve, one might try to produce a residual of the form

$$r_k = \chi_k(A)\varphi_k(A)r_0,$$

where φ_k is again the BiCG polynomial but χ_k is chosen to try and keep the residual norm small at each step while retaining the rapid overall convergence of CGS. For example, if $\chi_k(z)$ is of the form

$$(5.23) \qquad \chi_k(z) = (1 - \omega_k z)(1 - \omega_{k-1}z)\cdots(1 - \omega_1 z),$$

then the coefficients ω_j can be chosen at each step to minimize

$$\|r_j\| = \|(I - \omega_j A)\chi_{j-1}(A)\varphi_j(A)r_0\|.$$

This leads to the BiCGSTAB algorithm, which might be thought of as a combination of BiCG with Orthomin(1).

Again letting $\varphi_k(A)r_0$ denote the BiCG residual at step k and $\psi_k(A)r_0$ denote the BiCG direction vector at step k, recall that these polynomials satisfy recurrences (5.19–5.20). In the BiCGSTAB scheme we will need recurrences for

$$r_k = \chi_k(A)\varphi_k(A)r_0 \quad \text{and} \quad p_k = \chi_k(A)\psi_k(A)r_0.$$

It follows from (5.23) and (5.19–5.20) that

$$\begin{aligned} r_k &= (I - \omega_k A)\chi_{k-1}(A)[\varphi_{k-1}(A) - a_{k-1}A\psi_{k-1}(A)]r_0 \\ &= (I - \omega_k A)[r_{k-1} - a_{k-1}A p_{k-1}], \end{aligned}$$

$$\begin{aligned} p_k &= \chi_k(A)[\varphi_k(A) + b_k\psi_{k-1}(A)]r_0 \\ &= r_k + b_k(I - \omega_k A)p_{k-1}. \end{aligned}$$

Finally, we need to express the BiCG coefficients a_{k-1} and b_k in terms of the new vectors. Using the biorthogonality properties of the BiCG polynomials—$\langle \varphi_k(A)r_0, A^{H^j}\hat{r}_0 \rangle = \langle A\psi_k(A)r_0, A^{H^j}\hat{r}_0 \rangle = 0, \ j = 0, 1, \ldots, k-1$ (see Exercise 5.3)—together with the recurrence relations (5.19–5.20), we derive the following expressions for inner products appearing in the coefficient formulas (5.21–5.22):

$$\langle \varphi_{k-1}(A)r_0, \bar{\varphi}_{k-1}(A^H)\hat{r}_0 \rangle = (-1)^{k-1}a_{k-2} \cdots a_0 \langle \varphi_{k-1}(A)r_0, A^{H^{k-1}}\hat{r}_0 \rangle,$$

$$\langle A\psi_{k-1}(A)r_0, \bar{\psi}_{k-1}(A^H)\hat{r}_0 \rangle = (-1)^{k-1}a_{k-2} \cdots a_0 \langle A\psi_{k-1}(A)r_0, A^{H^{k-1}}\hat{r}_0 \rangle.$$

It also follows from these same biorthogonality and recurrence relations that the BiCGSTAB vectors satisfy

$$\begin{aligned} \langle r_{k-1}, \hat{r}_0 \rangle &= \langle \varphi_{k-1}(A)r_0, \bar{\chi}_{k-1}(A^H)\hat{r}_0 \rangle \\ &= (-1)^{k-1}\omega_{k-1} \cdots \omega_1 \langle \varphi_{k-1}(A)r_0, A^{H^{k-1}}\hat{r}_0 \rangle, \end{aligned}$$

$$\begin{aligned} \langle Ap_{k-1}, \hat{r}_0 \rangle &= \langle A\psi_{k-1}(A)r_0, \bar{\chi}_{k-1}(A^H)\hat{r}_0 \rangle \\ &= (-1)^{k-1}\omega_{k-1} \cdots \omega_1 \langle A\psi_{k-1}(A)r_0, A^{H^{k-1}}\hat{r}_0 \rangle, \end{aligned}$$

and hence the coefficient formulas (5.21–5.22) can be replaced by

$$a_{k-1} = \frac{\langle r_{k-1}, \hat{r}_0 \rangle}{\langle Ap_{k-1}, \hat{r}_0 \rangle}, \qquad b_k = \frac{a_{k-1}}{\omega_k} \frac{\langle r_k, \hat{r}_0 \rangle}{\langle r_{k-1}, \hat{r}_0 \rangle}.$$

This leads to the following algorithm.

Algorithm 6. BiCGSTAB.

Given x_0, compute $r_0 = b - Ax_0$ and set $p_0 = r_0$.
Choose $\hat{r}_0$ such that $\langle r_0, \hat{r}_0 \rangle \neq 0$. For $k = 1, 2, \ldots,$

Compute Ap_{k-1}.

Set $x_{k-1/2} = x_{k-1} + a_{k-1}p_{k-1}$, where $a_{k-1} = \frac{\langle r_{k-1}, \hat{r}_0 \rangle}{\langle Ap_{k-1}, \hat{r}_0 \rangle}$.

Compute $r_{k-1/2} = r_{k-1} - a_{k-1}Ap_{k-1}$.

Compute $Ar_{k-1/2}$.

Set $x_k = x_{k-1/2} + \omega_k r_{k-1/2}$, where $\omega_k = \frac{\langle r_{k-1/2}, Ar_{k-1/2} \rangle}{\langle Ar_{k-1/2}, Ar_{k-1/2} \rangle}$.

Compute $r_k = r_{k-1/2} - \omega_k Ar_{k-1/2}$.

Compute $p_k = r_k + b_k(p_{k-1} - \omega_k Ap_{k-1})$, where $b_k = \frac{a_{k-1}}{\omega_k} \frac{\langle r_k, \hat{r}_0 \rangle}{\langle r_{k-1}, \hat{r}_0 \rangle}$.

5.7. Which Method Should I Use?

For Hermitian problems, the choice of an iterative method is fairly straightforward—use CG or MINRES for positive definite problems and MIN-RES for indefinite problems. One can also use a form of the CG algorithm for indefinite problems. The relation between residual norms for CG and MINRES is like that for BiCG and QMR, however, as shown in Exercise 5.1. For this reason, the MINRES method is usually preferred. By using a simpler iteration, such as the simple iteration method described in section 2.1 or the Chebyshev method [94] which has not been described, one can avoid the inner products required in the CG and MINRES algorithms. This gives some savings in the cost of an iteration, but the price in terms of number of iterations usually outweighs the savings. An exception might be the case in which one has such a good preconditioner that even simple iteration requires only one or two steps. For some problems, multigrid methods provide such preconditioners.

The choice of an iterative method for non-Hermitian problems is not so easy. If matrix–vector multiplication is extremely expensive (e.g., if A is dense and has no special properties to enable fast matrix–vector multiplication), then (full) GMRES is probably the method of choice because it requires the fewest matrix–vector multiplications to reduce the residual norm to a desired level. If matrix–vector multiplication is not so expensive or if storage becomes a problem for full GMRES, then one of the methods described in this chapter is probably a good choice. Because of relation (5.17), we generally recommend QMR over BiCG.

The choice between QMR, CGS, and BiCGSTAB is problem dependent. There are also transpose-free versions of QMR that have not been described here [53]. Another approach, to be discussed in section 7.1, is to symmetrize the problem. For example, instead of solving $Ax = b$, one could solve $A^H Ax = A^H b$ or $AA^H y = b$ (so that $x = A^H y$) using the CG method. Of course, one does not actually form the normal equations; it is necessary only to compute matrix–vector products with A and A^H. How this approach compares with the methods described in this chapter is also problem dependent. In [102], the GMRES (full or restarted), CGS, and CGNE (CG for $AA^H y = b$) iterations were considered. For each method an example was constructed for which that method was by far the best and another example was given for which that method was by far the worst. Thus none of these methods can be eliminated as definitely inferior to one of the others, and none can be recommended as *the* method of choice for non-Hermitian problems.

To give some indication of the performance of the methods, we show here plots of residual norm and error norm versus the number of matrix–vector multiplications and versus the number of floating point operations (additions, subtractions, multiplications, and divisions) *assuming* that multiplication by A or A^H requires $9n$ operations. This is the cost of applying a 5-diagonal matrix A to a vector and probably is a lower bound on the cost of matrix–vector multiplication in most practical applications. The methods considered are full

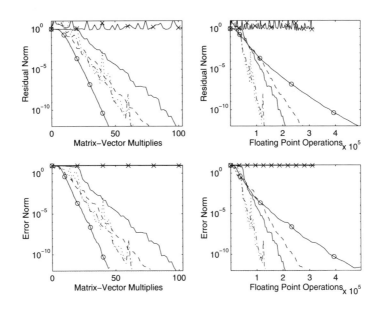

FIG. 5.2. *Performance of full GMRES (solid with o's), GMRES(10) (dashed), QMR (solid), CGS (dotted), BiCGSTAB (dash–dot), and CGNE (solid with x's).*

GMRES, GMRES(10) (that is, GMRES restarted after every 10 steps), QMR, CGS, BiCGSTAB, and CGNE.

The problem is the one described in section 5.4—a real 103-by-103 matrix A with random eigenvectors and with 50 pairs of complex conjugate eigenvalues randomly distributed in $[1, 2] \times [-i, i]$ and 3 additional real eigenvalues at 4, .5, and -1. Results are shown in Figure 5.2. The full GMRES algorithm necessarily requires the fewest matrix–vector multiplications to achieve a given residual norm. In terms of floating point operations, however, when matrix–vector multiplication requires only $9n$ operations, full GMRES is the most expensive method. The QMR algorithm uses two matrix–vector multiplications per step (one with A and one with A^H) to generate an approximation whose residual lies in the same Krylov space as the GMRES residual. Hence QMR requires at least twice as many matrix–vector multiplications to reduce the residual norm to a given level, and for this problem it requires only slightly more than this. A transpose-free variant of QMR would likely be more competitive. Since the CGS and BiCGSTAB methods construct a residual at step k that comes from the Krylov space of dimension $2k$ (using two matrix–vector multiplications per step), these methods could conceivably require as few matrix–vector multiplications as GMRES. For this example they require a moderate number of additional matrix–vector multiplications, but these seem to be the most efficient in terms of floating point operations. The CGNE method proved *very* inefficient for this problem, hardly reducing the error at all over the first 52 steps (104 matrix–

vector multiplications). The condition number of the matrix AA^H is 10^8, so this is not so surprising.

The results of this one test problem should *not* be construed as indicative of the relative performance of these algorithms for all or even most applications. In the Exercises, we give examples in which some of these methods perform far better or far worse than the others. It remains an open problem to characterize the classes of problems for which one method outperforms the others. For additional experimental results, see [128, 28].

Comments and Additional References.

The QMR algorithm was developed by Freund and Nachtigal [54]. Theorem 5.3.1 was given in [101].

Relation (5.13) in Lemma 5.4.1 has been established in a number of places (e.g., [22, 53, 54, 75, 111, 140]), but, surprisingly, the one-step leap to relation (5.14) and its consequence (5.17) seems not to have been taken explicitly until [29]. A similar relation between the GMRES and FOM residuals was observed in [22].

The CGS algorithm was developed by Sonneveld [124] and BiCGSTAB by van der Vorst [134].

An excellent survey article on iterative methods based on the nonsymmetric Lanczos algorithm, along with many references, is given in [78].

Exercises.

5.1. Use Lemma 5.4.1 to show that for Hermitian matrices A, the CG residual r_k^C is related to the MINRES residual r_k^M by

$$\|r_k^C\| = \frac{\|r_k^M\|}{\sqrt{1 - (\|r_k^M\|/\|r_{k-1}^M\|)^2}},$$

provided that the tridiagonal matrix T_k generated by the Lanczos algorithm is nonsingular.

5.2. Let r_k^Q denote the residual at step k of the QMR algorithm and let $T_{k+1,k}$ denote the $k+1$-by-k tridiagonal matrix generated by the Lanczos algorithm. Let T be any tridiagonal matrix whose upper left $k+1$-by-k block is $T_{k+1,k}$. Use the fact that the Arnoldi algorithm, applied to T with initial vector ξ_1, generates the same matrix $T_{k+1,k}$ at step k to show that

$$\|r_k^Q\| \leq \|V_{k+1}\| \cdot \|r_k^G(T)\| \leq \sqrt{k+1}\, \|r_k^G(T)\|,$$

where $r_k^G(T)$ is the residual at step k of the GMRES algorithm applied to the linear system $T\chi = \|r_0\|\xi_1$. If T is taken to be the tridiagonal matrix generated at step n of the Lanczos algorithm (assuming the algorithm does not break down or terminate before step n), then the eigenvalues of T are the same as those of A. Thus the convergence of QMR is like that of GMRES applied to a matrix with the same eigenvalues as A. This does not provide useful a priori information about the convergence rate of QMR, however, as it was noted in section 3.2 that eigenvalue information alone tells nothing about the behavior of GMRES. (This result was proved in [54] for the special case $T = T_{k+1}$, and it was proved in [28] for $T = T_n$.)

5.3. Prove the biconjugacy relations

$$\langle r_k, A^{H^j} \hat{r}_0 \rangle = \langle A p_k, A^{H^j} \hat{r}_0 \rangle = 0, \quad j < k$$

for the BiCG algorithm.

5.4. The following examples are taken from [102]. They demonstrate that the performance of various iterative methods can differ dramatically for a given problem and that the best method for one problem may be the worst for another.

(a) *CGNE wins.* Suppose A is the unitary shift matrix

$$\begin{pmatrix} 0 & 1 & 0 & \cdots & 0 \\ & & 1 & \cdots & 0 \\ & & & \ddots & \\ & & & & 1 \\ 1 & 0 & \cdots & \cdots & 0 \end{pmatrix}$$

and b is the first unit vector ξ_1. How many iterations will the full GMRES method need to solve $Ax = b$, with a zero initial guess? What is a lower bound on the number of matrix–vector multiplications required by CGS? How many iterations are required if one applies CG to the normal equations $AA^H y = b$, $x = A^H y$?

(b) *CGNE loses.* Suppose A is the block diagonal matrix

$$\begin{pmatrix} M_1 & & & \\ & M_2 & & \\ & & \ddots & \\ & & & M_{n/2} \end{pmatrix}, \quad M_j = \begin{pmatrix} 1 & j-1 \\ 0 & 1 \end{pmatrix}, \quad 1 \le j \le n/2.$$

What is the degree of the minimal polynomial of A? How many steps will GMRES require to obtain the solution to a linear system $Ax = b$? How many matrix–vector multiplications will CGS require, assuming that $\hat{r}_0 = r_0$? The singular values of this matrix lie approximately in the range $[2/n, n/2]$. Would you expect CGNE to require few or many iterations if n is large?

(c) *CGS wins.* Suppose A is a Hermitian matrix with many eigenvalues distributed throughout an interval $[c, d]$ on the positive real axis. Which method would you expect to require the least amount of *work* to solve a linear system $Ax = b$—(full) GMRES, CGS, or CGNE? Explain your answer. (Of course, one would do better to solve a Hermitian problem using CG or MINRES, but perhaps it is not known that A is Hermitian.)

(d) *CGS loses.* Let A be the skew-symmetric matrix

$$\begin{pmatrix} 0 & 1 \\ -1 & 0 \end{pmatrix} \times I_{n/2},$$

that is, an n-by-n block diagonal matrix with 2-by-2 blocks. Show that this matrix is normal and has eigenvalues $\pm \imath$ and singular value 1. How many steps are required to solve a linear system $Ax = b$ using CGNE? GMRES? Show, however, that for any real initial residual r_0, if $\hat{r}_0 = r_0$ then CGS breaks down with a division by 0 at the first step.

5.5. When the two-sided Lanczos algorithm is used in the solution of linear systems, the right starting vector is always the initial residual, $r_0/\|r_0\|$, but the left starting vector $\hat{r}_0$ is not specified. Consider an *arbitrary* 3-term recurrence:

$$\tilde{v}_{j+1} = Av_j - \alpha_j v_j - \beta_{j-1} v_{j-1},$$

$$v_{j+1} = \tilde{v}_{j+1}/\gamma_j, \quad \text{where } \gamma_j = \|\tilde{v}_{j+1}\|.$$

The γ's are chosen so that the vectors have norm 1, but the α's and β's can be anything. Show that if this recurrence is run for no more than $[(n+2)/2]$ steps, then there is a nonzero vector w_1 such that

$$(5.24) \quad \langle v_j, A^{H^\ell} w_1 \rangle = 0 \quad \forall \ell < j - 1, \quad j = 2, \ldots, \left[\frac{n+2}{2}\right];$$

i.e., assuming there is no exact breakdown with $\langle v_j, w_j \rangle = 0$, the arbitrary recurrence *is* the two-sided Lanczos algorithm for a certain left starting vector w_1. (Hint: The condition (5.24) is equivalent to $\langle w_1, A^\ell v_j \rangle = 0$ $\forall \ell < j - 1, j = 2, \ldots, [(n+2)/2]$. Show that there are only $n-1$ linearly independent vectors to which w_1 must be orthogonal.)

This somewhat disturbing result suggests that some assumptions must be made about the left starting vector, if we are to have any hope of establishing good a priori error bounds for the Lanczos-based linear system solvers [67]. In practice, however, it is observed that the convergence behavior of these methods is about the same for most randomly chosen left starting vectors or for $\hat{r}_0 = r_0$, which is sometimes recommended.

Chapter 6

Is There a Short Recurrence for a Near-Optimal Approximation?

Of the many non-Hermitian iterative methods described in the previous chapter, none can be shown to generate a near-optimal approximate solution for every initial guess. It sometimes happens that the QMR approximation at step k is almost as good as the (optimal) GMRES approximation, but sometimes this is not the case. It was shown by Faber and Manteuffel [45] that if "optimal" is taken to mean having the smallest possible error in some inner product norm that is *independent* of the initial vector, then the optimal approximation cannot be generated with a short recurrence. The details of this result are provided in section 6.1. The result should not necessarily be construed as ruling out the possibility of a clear "method of choice" for non-Hermitian problems. Instead, it may suggest directions in the search for such a method. Possibilities are discussed in section 6.2.

6.1. The Faber and Manteuffel Result.

Consider a recurrence of the following form. Given x_0, compute $p_0 = b - Ax_0$, and for $k = 1, 2, \ldots$, set

$$(6.1) \qquad x_k = x_{k-1} + a_{k-1}p_{k-1},$$

$$(6.2) \qquad p_k = Ap_{k-1} - \sum_{j=k-s+1}^{k-1} b_{k-1,j}p_j$$

for some coefficients a_{k-1} and $b_{k-1,j}$, $j = k - s + 1, \ldots, k - 1$, where s is some integer less than n. It is easy to show by induction that the approximate solution x_k generated by this recurrence is of the form

$$(6.3) \qquad x_k \in x_0 + \text{span}\{r_0, Ar_0, \ldots, A^{k-1}r_0\}$$

and that the direction vectors $p_0, \ldots, p_{k-1}$ form a basis for the Krylov space

$$\text{span}\{p_0, p_1, \ldots p_{k-1}\} = \text{span}\{r_0, Ar_0, \ldots, A^{k-1}r_0\}.$$

The recurrence Orthodir(3) is of the form (6.1–6.2), with $s = 3$, as is Orthomin(2). To see that Orthomin(2) is of this form, note that in that algorithm we have

$$p_k = r_k - b_{k-1}p_{k-1}, \qquad \text{where } r_k = r_{k-1} - a_{k-1}Ap_{k-1}.$$

Substituting for r_k in the recurrence for p_k gives

$$p_k = r_{k-1} - a_{k-1}Ap_{k-1} - b_{k-1}p_{k-1},$$

and using the fact that $r_{k-1} = p_{k-1} + b_{k-2}p_{k-2}$ gives

$$p_k = -a_{k-1}Ap_{k-1} + (1 - b_{k-1})p_{k-1} + b_{k-2}p_{k-2}.$$

The normalization of p_k is of no concern, since that can be accounted for by choosing the coefficient a_{k-1} appropriately, so if p_k is replaced by $[(-1)^k \prod_{j=1}^{k-1} a_j^{-1}]p_k$, then Orthomin(2) fits the pattern (6.1–6.2). The MINRES algorithm for Hermitian problems is also of this form, and it has the desirable property of generating, at each step, the approximation of the form (6.3) for which the 2-norm of the residual is minimal. The CG algorithm for Hermitian positive definite problems is also of the form (6.1–6.2), and at each step it generates the approximation of the form (6.3) for which the A-norm of the error is minimal.

For what matrices A can one construct a recurrence of the form (6.1–6.2) with the property that for any initial vector x_0, the approximation x_k at step k is the "optimal" approximation from the space (6.3), where "optimal" means that the error $e_k \equiv A^{-1}b - x_k$ is minimal in some *inner product* norm, the inner product being *independent* of the initial vector? This is essentially the question answered by Faber and Manteuffel [45]. See also [46, 5, 86, 138]. We will not include the entire proof, but the answer is that for $s < \sqrt{n}$, except for a few anomalies, the matrices for which such a recurrence exists are those of the form $B^{-1/2}CB^{1/2}$, where C is either Hermitian or of the form $C = e^{i\theta}(dI + F)$, with d real and $F^H = -F$, and B is a Hermitian positive definite matrix. Equivalently (Exercise 6.1), such a recurrence exists for matrices A of the form

$$(6.4) \qquad A = e^{i\phi}(icI + G), \quad c \geq 0, \quad 0 \leq \phi \leq 2\pi, \quad B^{-1}G^H B = G.$$

If A is of the form (6.4), then $B^{1/2}AB^{-1/2}$ is just a shifted and rotated Hermitian matrix.

To see why this class of matrices is special, note that the error e_k in a recurrence of the form (6.1–6.2) satisfies

$$(6.5) \qquad e_k \in e_0 + \text{span}\{p_0, p_1, \ldots, p_{k-1}\}.$$

If $\langle\langle \cdot, \cdot \rangle\rangle$ denotes the inner product in which the norm of e_k is minimized, then e_k must be the unique vector of the form (6.5) satisfying

$$\langle\langle e_k, p_j \rangle\rangle = 0, \quad j = 0, 1, \ldots, k - 1.$$

It follows that since $e_k = e_{k-1} - a_{k-1}p_{k-1}$, the coefficient a_{k-1} must be

$$a_{k-1} = \frac{\langle\langle e_{k-1}, p_{k-1}\rangle\rangle}{\langle\langle p_{k-1}, p_{k-1}\rangle\rangle}.$$

For $j < k-1$, we have

$$\langle\langle e_k, p_j\rangle\rangle = \langle\langle e_{k-1}, p_j\rangle\rangle - a_{k-1}\langle\langle p_{k-1}, p_j\rangle\rangle,$$

so if $\langle\langle e_{k-1}, p_j\rangle\rangle = 0$, then in order to have $\langle\langle e_k, p_j\rangle\rangle = 0$, it is necessary that either $a_{k-1} = 0$ or $\langle\langle p_{k-1}, p_j\rangle\rangle = 0$. If it is required that $\langle\langle p_k, p_j\rangle\rangle = 0$ for all k and all $j < k$, then the coefficients $b_{k-1,j}$ must be given by

$$b_{k-1,j} = \frac{\langle\langle Ap_{k-1}, p_j\rangle\rangle}{\langle\langle p_j, p_j\rangle\rangle}.$$

A precise statement of the Faber and Manteuffel result is given in the following definition and theorem.

DEFINITION 6.1.1. *An algorithm of the form (6.1–6.2) is an s-term CG method for A if, for every p_0, the vectors p_k, $k = 1, 2, \ldots, m - 1$, satisfy $\langle\langle p_k, p_j\rangle\rangle = 0$ for all $j < k$, where m is the number of steps required to obtain the exact solution $x_m = A^{-1}b$.*

THEOREM 6.1.1 (Faber and Manteuffel [45]). *An s-term CG method exists for the matrix A if and only if either*

(i) *the minimal polynomial of A has degree less than or equal to s, or*

(ii) *$A^\star$ is a polynomial of degree less than or equal to $s - 2$ in A, where $A^\star$ is the adjoint of A with respect to some inner product, that is, $\langle\langle Av, w\rangle\rangle = \langle\langle v, A^\star w\rangle\rangle$ for all vectors v and w.*

Proof (of sufficiency only). The choice of coefficients a_i and $b_{i,j}$, $i = 0, \ldots, s - 1$, $j \leq i$ not only forces $\langle\langle p_k, p_j\rangle\rangle = 0$, $k = 1, \ldots, s - 1$, $j < k$, but also ensures that the error at steps 1 through s is minimized in the norm corresponding to the given inner product. Since the error at step k is equal to a certain kth-degree polynomial in A times the initial error, if the minimal polynomial of A has degree $k \leq s$, then the algorithm will discover this minimal polynomial (or another one for which $e_k = p_k(A)e_0 = 0$), and the exact solution will be obtained after $k \leq s$ steps. In this case, then, iteration (6.1–6.2) is an s-term CG method.

For $k \geq s$ and $i < k - s + 1$, it follows from (6.2) that

$$\langle\langle p_k, p_i\rangle\rangle = \langle\langle Ap_{k-1}, p_i\rangle\rangle - \sum_{j=k-s+1}^{k-1} b_{k-1,j}\langle\langle p_j, p_i\rangle\rangle.$$

If $\langle\langle p_j, p_i\rangle\rangle = 0$ for $j = k - s + 1, \ldots, k - 1$, then we will have $\langle\langle p_k, p_i\rangle\rangle = 0$ if and only if

(6.6) $\langle\langle Ap_{k-1}, p_i \rangle\rangle \equiv \langle\langle p_{k-1}, A^\star p_i \rangle\rangle = 0.$

If $A^\star = q_{s-2}(A)$ for some polynomial q_{s-2} of degree $s-2$ or less, then (6.6) will hold, since p_{k-1} is orthogonal to the space

$$\text{span}\{p_0, \ldots, p_{k-2}\} = \text{span}\{p_0, Ap_0, \ldots, A^{k-2}p_0\},$$

which contains $q_{s-2}(A)p_i$ since $i + s - 2 \leq k - 2$. □

To clarify condition (ii) in Theorem 6.1.1, first recall (section 1.3.1) that for any inner product $\langle\langle \cdot, \cdot \rangle\rangle$ there is a Hermitian positive definite matrix B such that

$$\langle\langle v, w \rangle\rangle = \langle v, Bw \rangle$$

for all vectors v and w, where $\langle \cdot, \cdot \rangle$ denotes the standard Euclidean inner product. The B-adjoint of A, denoted $A^\star$ in the theorem, is the unique matrix satisfying

$$\langle Av, Bw \rangle = \langle v, BA^\star w \rangle$$

for all v and w. From this definition it follows that

$$A^\star = B^{-1}A^H B,$$

where the superscript ^{H} denotes the adjoint in the Euclidean norm $A^H = \bar{A}^T$. The matrix A is said to be B-normal if and only if $A^\star A = AA^\star$. If $B^{1/2}$ denotes the Hermitian positive definite square root of B, then this is equivalent to the condition that

$$(B^{-1/2}A^H B^{1/2})(B^{1/2}AB^{-1/2}) = (B^{1/2}AB^{-1/2})(B^{-1/2}A^H B^{1/2}),$$

which is the condition that $B^{1/2}AB^{-1/2}$ be normal.

Let B be fixed and let $\tilde{A}$ denote the matrix $B^{1/2}AB^{-1/2}$. It can be shown (Exercise 6.2) that $\tilde{A}$ is normal (A is B-normal) if and only if $\tilde{A}^H$ can be written as a polynomial (of some degree) in $\tilde{A}$. If η is the smallest degree for which this is true, then η is called the B-normal degree of A. For any integer $t \geq \eta$, A is said to be B-normal(t). With this notation, condition (ii) of Theorem 6.1.1 can be stated as follows:

(ii′) A is B-normal($s - 2$).

Condition (ii′) still may seem obscure, but the following theorem, also from [45], shows that matrices A with B-normal degree η greater than 1 but less than $\sqrt{n}$ also have minimal polynomials of degree less than n. These matrices belong to a subspace of $\mathbf{C}^{n \times n}$ of dimension less than n^2, so they might just be considered anomalies. The more interesting case is $\eta = 1$ or the B-normal(1) matrices in (ii′).

THEOREM 6.1.2 (Faber and Manteuffel [45]). *If A has B-normal degree $\eta > 1$, then the minimal polynomial of A has degree less than or equal to η^2.*

Proof. The degree $d(A)$ of the minimal polynomial of A is the same as that of $\tilde{A} \equiv B^{1/2}AB^{-1/2}$. Since $\tilde{A}$ is normal, it has exactly $d(A)$ distinct eigenvalues, and we will have $\tilde{A}^H = q(A)$ if and only if

$$q(\lambda_i) = \bar{\lambda}_i, \quad i = 1, \ldots, d(A).$$

How many distinct complex numbers z can satisfy $q(z) = \bar{z}$? Note that $\bar{q}(\bar{z}) = z$ or $\bar{q}(q(z)) = z$. The expression $\bar{q}(q(z)) - z$ is a polynomial of degree exactly η^2 if q has degree $\eta > 1$. (If the degree of q were 1, this expression could be identically zero.) It follows that there are at most η^2 distinct roots, so $d(A) \leq \eta^2$.   □

The B-normal(1) matrices, for which a 3-term CG method exists, are characterized in the following theorem.

THEOREM 6.1.3 (Faber and Manteuffel [45]).   *If A is B-normal(1) then $d(A) = 1$, $A^\star = A$, or*

$$\tilde{A} \equiv B^{1/2}AB^{-1/2} = e^{\imath\theta}\left(\frac{r}{2}I + F\right),$$

where r is real and $F = -F^H$.

Proof. Since $\tilde{A}$ is normal, if $\tilde{A}$ has all real eigenvalues, then $\tilde{A}^H = \tilde{A}$ or $A^\star = A$.

Suppose $\tilde{A}$ has at least one complex eigenvalue.   There is a linear polynomial q such that each of the eigenvalues λ_i of $\tilde{A}$ satisfies $q(\lambda_i) = \bar{\lambda}_i$. This implies that $\bar{q}(\bar{\lambda}_i) = \lambda_i$ or $\bar{q}(q(\lambda_i)) - \lambda_i = 0$. In general, this equation has just one root λ_i, and if this is the case then $d(A) = 1$.

Let $q(z) = az - b$. The expression $\bar{q}(q(\lambda_i)) - \lambda_i = 0$ can be written as

$$(\bar{a}a - 1)\lambda_i - (\bar{a}b + \bar{b}) = 0.$$

There is more than one root λ_i only if the expression on the left is identically zero, which means that $a = -b/\bar{b}$. Let $b = re^{\imath\theta}$, $\imath \equiv \sqrt{-1}$. Then

$$q(z) = -e^{\imath\theta}(ze^{-\imath\theta} - r).$$

If $q(z) = \bar{z}$, then

$$-(ze^{-\imath\theta} - r) = \bar{z}e^{\imath\theta} = \overline{ze^{-\imath\theta}},$$

which yields

$$r = ze^{-\imath\theta} + \overline{ze^{-\imath\theta}}.$$

Thus, if λ is an eigenvalue of $\tilde{A}$, the real part of $\lambda e^{-\imath\theta}$ is $r/2$. This implies that

$$F = e^{-\imath\theta}\tilde{A} - \frac{r}{2}I$$

has only pure imaginary eigenvalues; hence, since $\tilde{A}$ is normal, $F = -F^H$.   □

6.2. Implications.

The class of B-normal(1) matrices of the previous section are matrices for which CG methods are already known. They are diagonalizable matrices whose spectrum is contained in a line segment in the complex plane. See [26, 142].

Theorems 6.1.1–6.1.3 imply that for most non-Hermitian problems, one cannot expect to find a short recurrence that generates the optimal approximation from successive Krylov spaces, if "optimality" is defined in terms of an inner product norm that is independent of the initial vector. It turns out that most non-Hermitian iterative methods actually do find the optimal approximation in some norm [11] (see Exercise 6.3). Unfortunately, however, it is a norm that cannot be related easily to the 2-norm or the ∞-norm or any other norm that is likely to be of interest. For example, the BiCG approximation is optimal in the $P_n^{-H} P_n^{-1}$-norm, where the columns of P_n are the biconjugate direction vectors. The QMR approximation is optimal in the $A^H V_n^{-H} V_n^{-1} A$-norm, where the columns of V_n are the biorthogonal basis vectors.

The possibility of a short recurrence that would generate optimal approximations in some norm that depends on the initial vector but that can be shown to differ from, say, the 2-norm by no more than some moderate size factor remains. This might be the best hope for developing a clear "method of choice" for non-Hermitian linear systems.

It should also be noted that the Faber and Manteuffel result deals only with a *single* recurrence. It is still an open question whether coupled short recurrences can generate optimal approximations. For some preliminary results, see [12].

It remains a major open problem to find a method that generates provably "near-optimal" approximations in some standard norm while still requiring only $O(n)$ work and storage (in addition to the matrix–vector multiplication) at each iteration—or to prove that such a method does not exist.

Exercises.

6.1. Show that a matrix A is of the form (6.4) if and only if it is of the form $B^{-1/2}CB^{1/2}$, where C is either Hermitian or of the form $e^{i\theta}(dI + F)$, with d real and $F^H = -F$.

6.2. Show that a matrix A is normal if and only if $A^H = q(A)$ for some polynomial q. (Hint: If A is normal, write A in the from $A = U\Lambda U^H$, where Λ is diagonal and U is unitary, and determine a polynomial q for which $q(\Lambda) = \bar{\Lambda}$.)

6.3. The following are special instances of results due to Barth and Manteuffel [11]:

(a) Assume that the BiCG iteration does not break down or find the exact solution before step n. Use the fact that the BiCG error at step k is of the form

$$e_k \in e_0 + \text{span}\{p_0, p_1, \ldots, p_{k-1}\}$$

and the residual satisfies

$$r_k \perp \text{span}\{\hat{p}_0, \hat{p}_1, \ldots, \hat{p}_{k-1}\}$$

to show that the BiCG approximation at each step is optimal in the $P_n^{-H}P_n^{-1}$-norm, where the columns of P_n are the biconjugate direction vectors.

(b) Assume that the two-sided Lanczos recurrence does not break down or terminate before step n. Use the fact that the QMR error at step k is of the form $e_k = e_0 - V_k y_k$ and that the QMR residual satisfies $r_k^H V_{k+1} V_{k+1}^H A V_k = 0$ to show that the QMR approximation at each step is optimal in the $A^H V_n^{-H} V_n^{-1} A$-norm, where the columns of V_n are the biorthogonal basis vectors.

6.4. Write down a CG method for matrices of the form $I - F$, where $F = -F^H$, which minimizes the 2-norm of the residual at each step. (Hint: Note that one can use a 3-term recurrence to construct an orthonormal basis for the Krylov space $\text{span}\{q_1, (I - F)q_1, \ldots, (I - F)^{k-1}q_1\}$, when F is skew-Hermitian.)

Chapter 7

Miscellaneous Issues

7.1. Symmetrizing the Problem.

Because of the difficulties in solving non-Hermitian linear systems, one might consider converting a non-Hermitian problem to a Hermitian one by solving the *normal equations*. That is, one applies an iterative method to one of the linear systems

$$(7.1) \qquad A^H A x = A^H b \quad \text{or} \quad A A^H y = b, \quad x = A^H y.$$

As usual, this can be accomplished without actually forming the matrices $A^H A$ or $A A^H$, and, in the latter case, one need not explicitly generate approximations to y but instead can carry along approximations $x_k \equiv A^H y_k$. For instance, if the CG method is used to solve either of the systems in (7.1), then the algorithms, sometimes called CGNR and CGNE, respectively, can be implemented as follows.

Algorithm 7. CG for the Normal Equations (CGNR and CGNE).

Given an initial guess x_0, compute $r_0 = b - A x_0$.
Compute $A^H r_0$ and set $p_0 = A^H r_0$. For $k = 1, 2, \ldots$,

Compute $A p_{k-1}$.

Set $x_k = x_{k-1} + a_{k-1} p_{k-1}$, where
$$a_{k-1} = \frac{\langle A^H r_{k-1}, A^H r_{k-1} \rangle}{\langle A p_{k-1}, A p_{k-1} \rangle} \text{ for CGNR}, \quad a_{k-1} = \frac{\langle r_{k-1}, r_{k-1} \rangle}{\langle p_{k-1}, p_{k-1} \rangle} \text{ for CGNE}.$$

Compute $r_k = r_{k-1} - a_{k-1} A p_{k-1}$.

Compute $A^H r_k$.

Set $p_k = A^H r_k + b_{k-1} p_{k-1}$, where
$$b_{k-1} = \frac{\langle A^H r_k, A^H r_k \rangle}{\langle A^H r_{k-1}, A^H r_{k-1} \rangle} \text{ for CGNR}, \quad b_{k-1} = \frac{\langle r_k, r_k \rangle}{\langle r_{k-1}, r_{k-1} \rangle} \text{ for CGNE}.$$

The CGNR algorithm minimizes the $A^H A$-norm of the error, which is the

2-norm of the residual $b - Ax_k$, over the affine space

$$x_k \in x_0 + \text{span}[A^H r_0, (A^H A)A^H r_0, \ldots, (A^H A)^{k-1} A^H r_0].$$

The CGNE algorithm minimizes the AA^H-norm of the error in y_k, which is the 2-norm of the error $x - x_k$, over the affine space

$$x_k \in x_0 + \text{span}[A^H r_0, A^H (AA^H)r_0, \ldots, A^H (AA^H)^{k-1} r_0].$$

Note that these two spaces are the same, and both involve powers of the symmetrized matrix $A^H A$ or AA^H.

Numerical analysts sometimes *cringe* at the thought of solving the normal equations for two reasons. First, since the condition number of $A^H A$ or AA^H is the *square* of the condition number of A, if there were an iterative method for solving $Ax = b$ whose convergence rate was governed by the condition number of A, then squaring this condition number would significantly degrade the convergence rate. Unfortunately, however, for a non-Hermitian matrix A, there is no iterative method whose convergence rate is governed by the condition number of A.

The other objection to solving the normal equations is that one cannot expect to achieve as high a level of accuracy when solving a linear system $Cy = d$ as when solving $Ax = b$, if the condition number of C is greater than that of A. This statement can be based on a simple perturbation argument. Since the entries of C and d probably cannot be represented exactly on the computer (or in the case of iterative methods, the product of C with a given vector cannot be computed exactly), the best approximation $\tilde{y}$ to y that one can hope to find numerically is one that satisfies a nearby system $(C + \delta C)\tilde{y} = d + \delta d$, where the size of δC and δd are determined by the machine precision. If $\tilde{y}$ is the solution of such a perturbed system, then it can be shown, for sufficiently small perturbations, that

$$\frac{\|y - \tilde{y}\|}{\|y\|} \leq \kappa(C) \left(\frac{\|\delta C\|}{\|C\|} + \frac{\|\delta d\|}{\|d\|} \right).$$

If C and d were the only data available for the problem, then the terms $\|\delta C\|/\|C\|$ and $\|\delta d\|/\|d\|$ on the right-hand side of this inequality could not be expected to be less than about ϵ, the machine precision. For the CGNR and CGNE methods, however, not only is $C = A^H A$ available, but the matrix A itself is available. (That is, one can apply A to a given vector.) As a result, it will be shown in section 7.4 that the achievable level of accuracy is about the same as that for the original linear system.

Thus, neither of the standard arguments against solving the normal equations is convincing. There are problems for which the CGNR and CGNE methods are best (Exercise 5.4a), and there are other problems for which one of the non-Hermitian matrix iterations far outperforms these two (Exercise 5.4b). In practice, the latter situation seems to be more common. There is little theory characterizing problems for which the normal equations approach is or is not to be preferred to a non-Hermitian iterative method. (See Exercise 7.1, however.)

7.2. Error Estimation and Stopping Criteria.

When a linear system is "solved" by a computer using floating point arithmetic, one does not obtain the exact solution, whether a direct or an iterative method is employed. With iterative methods especially, it is important to have some idea of what constitutes an acceptably good approximate solution, so the iteration can be stopped when this level of accuracy is achieved. Often it is desired to have an approximate solution for which some standard norm of the error, say, the 2-norm or the ∞-norm, is less than some tolerance. One can compute the residual $b - Ax_k$, but one cannot compute the error $A^{-1}b - x_k$. Hence one might try to estimate the desired error norm using the residual or other quantities generated during the iteration.

The relative error norm is related to the relative residual norm by

$$(7.2) \qquad \frac{1}{\kappa(A)} \frac{|||b - Ax_k|||}{|||b|||} \leq \frac{|||A^{-1}b - x_k|||}{|||A^{-1}b|||} \leq \kappa(A)\frac{|||b - Ax_k|||}{|||b|||},$$

where $\kappa(A) = |||A||| \cdot |||A^{-1}|||$ and $||| \cdot |||$ represents any vector norm and its induced matrix norm. To see this, note that since $b - Ax_k = A(A^{-1}b - x_k)$, we have

$$(7.3) \qquad |||b - Ax_k||| \leq |||A||| \cdot |||A^{-1}b - x_k|||,$$

$$(7.4) \qquad |||A^{-1}b - x_k||| \leq |||A^{-1}||| \cdot |||b - Ax_k|||.$$

Since we also have $|||A^{-1}b||| \leq |||A^{-1}||| \cdot |||b|||$, combining this with (7.3) gives the first inequality in (7.2). Using the inequality $|||b||| \leq |||A||| \cdot |||A^{-1}b|||$ with (7.4) gives the second inequality in (7.2). To obtain upper and lower bounds on the desired error norm, one might therefore attempt to estimate the condition number of A in this norm.

It was noted at the end of Chapter 4 that the eigenvalues of the tridiagonal matrix T_k generated by the Lanczos algorithm (and, implicitly, by the CG and MINRES algorithms for Hermitian matrices) provide estimates of some of the eigenvalues of A. Hence, if A is Hermitian and the norm in (7.2) is the 2-norm, then the eigenvalues of T_k can be used to estimate $\kappa(A)$. It is easy to show (Exercise 7.2) that the ratio of largest to smallest eigenvalue of T_k gives a lower bound on the condition number of A, but in practice it is usually a very good estimate, even for moderate size values of k. Hence one might stop the iteration when

$$(7.5) \qquad \kappa(T_k)\frac{\|b - Ax_k\|}{\|b\|} \leq \text{tol},$$

where tol is the desired tolerance for the 2-norm of the relative error. This could cause the iteration to terminate too soon, since $\kappa(T_k) \leq \kappa(A)$, but, more often, it results in extra iterations because the right-hand side of (7.2) is an overestimate of the actual error norm.

Unfortunately, for most other norms in (7.2), the condition number of A cannot be well approximated by the condition number (or any other simple function) of T_k. For non-Hermitian matrix iterations such as BiCG, QMR, and GMRES, the condition number of A cannot be approximated easily using the underlying non-Hermitian tridiagonal or upper Hessenberg matrix. (In the GMRES algorithm, one might consider approximating $\kappa(A)$ in the 2-norm by the ratio of largest to smallest singular value of H_k, since, at least for $k = n$, we have $\kappa(H_n) = \kappa(A)$. Unfortunately, however, for $k < n$, the singular values of H_k do not usually provide good estimates of the singular values of A.)

Since the right-hand side of (7.2) is an overestimate of the actual error norm, one might try to use quantities generated during the iteration in different ways. Sometimes an iteration is stopped when the difference between two consecutive iterates or between several consecutive iterates is less than some tolerance. For most iterative methods, however, this could lead to termination before the desired level of accuracy is achieved.

Consider the CG algorithm where A is Hermitian and positive definite and it is desired to reduce the A-norm of the error to a certain tolerance. The error $e_k \equiv x - x_k$ satisfies

$$e_k = e_{k-1} - a_{k-1}p_{k-1},$$

$$(7.6) \quad \langle e_k, Ae_k \rangle = \langle e_{k-1}, Ae_{k-1} \rangle - |a_{k-1}\langle r_{k-1}, p_{k-1} \rangle|$$

$$(7.7) \quad = \langle e_{k-d}, Ae_{k-d} \rangle - \sum_{j=1}^{d} |a_{k-j}\langle r_{k-j}, p_{k-j} \rangle|.$$

The A-norm of the difference $x_k - x_{k-1} = a_{k-1}p_{k-1}$ gives a *lower bound* on the error at step $k - 1$:

$$\langle e_{k-1}, Ae_{k-1} \rangle \geq |a_{k-1}\langle r_{k-1}, p_{k-1} \rangle| = \|x_k - x_{k-1}\|_A^2.$$

To obtain an upper bound, one can use the fact that the A-norm of the error is reduced by at least the factor $(\kappa - 1)/(\kappa + 1)$ at every step, or, more generally, that the A-norm of the error is reduced by at least the factor

$$(7.8) \quad \gamma_d \equiv 2 \left[\left(\frac{\sqrt{\kappa} - 1}{\sqrt{\kappa} + 1} \right)^d + \left(\frac{\sqrt{\kappa} + 1}{\sqrt{\kappa} - 1} \right)^d \right]^{-1}$$

after every d steps. Here κ denotes the condition number of A in the 2-norm and is the ratio of largest to smallest eigenvalue of A. This error bound follows from the fact that the CG polynomial is optimal for minimizing the A-norm of the error and hence is at least as good as the product of the $(k-d)$th-degree CG polynomial with the dth-degree Chebyshev polynomial. (See Theorem 3.1.1.)

Using (7.8) with (7.7), we find

$$(7.9) \qquad \langle e_{k-d}, A e_{k-d} \rangle \leq \left(\frac{1}{1 - \gamma_d^2} \right) \sum_{j=1}^{d} |a_{k-j} \langle r_{k-j}, p_{k-j} \rangle |.$$

One could again estimate κ using the ratio of largest to smallest eigenvalue of T_k and stop when the quantity in (7.9) is less than a given tolerance for some value of d.

Since (7.9) still involves an upper bound on the error which, for some problems, is not a very good estimate, different approaches have been considered. One of the more interesting ones involves looking at the quadratic form $r_k^H A^{-1} r_k$ as an integral and using different quadrature formulas to obtain upper and lower bounds for this integral [57]. The bounds obtained in this way appear to give very good estimates of the actual A-norm of the error. The subject of effective stopping criteria for iterative methods remains a topic of current research.

7.3. Attainable Accuracy.

Usually, the accuracy required from an iterative method is considerably less than it is capable of ultimately achieving. The important property of the method is the number of iterations or total work required to achieve a fairly modest level of accuracy. Occasionally, however, iterative methods are used for very ill-conditioned problems, and then it is important to know how the machine precision and the condition number of the matrix limit the attainable accuracy. Such analysis has been carried out for a number of iterative methods, and here we describe some results for a class of methods which includes the CG algorithm (Algorithm 2), the CGNR and CGNE algorithms (Algorithm 7), and some implementations of the MINRES, BiCG, and CGS algorithms (although not the ones recommended here). We will see in Part II of this book that the preconditioned versions of these algorithms also fall into this category.

The analysis applies to algorithms in which the residual vector r_k is updated rather than computed directly, using formulas of the form

$$(7.10) \qquad x_k = x_{k-1} + a_{k-1} p_{k-1}, \quad r_k = r_{k-1} - a_{k-1} A p_{k-1}.$$

Here p_{k-1} is some direction vector and a_{k-1} is some coefficient. It is assumed that the initial residual is computed directly as $r_0 = b - A x_0$.

It can be shown that when formulas (7.10) are implemented in finite precision arithmetic, the difference between the true residual $b - A x_k$ and the updated vector r_k satisfies

$$(7.11) \qquad \frac{\|b - A x_k - r_k\|}{\|A\| \, \|x\|} \leq \epsilon \, O(k) \left(1 + \max_{j \leq k} \frac{\|x_j\|}{\|x\|} \right),$$

where ϵ is the machine precision [66]. The growth in intermediate iterates, reflected on the right-hand side of (7.11), appears to play an important role in determining the size of the quantity on the left-hand side.

It is often observed numerically (but in most cases has *not* been proved) that the vectors r_k converge to zero as $k \to \infty$ or, at least, that their norms become many orders of magnitude smaller than the machine precision. In such cases, the right-hand side of (7.11) (without the $O(k)$ factor, which is an overestimate) gives a reasonable estimate of the best attainable actual residual:

$$(7.12) \qquad \min_k \frac{\|b - Ax_k\|}{\|A\| \, \|x\|} \approx c \, \epsilon \, \max_k \frac{\|x_k\|}{\|x\|},$$

where c is a moderate size constant.

The quantity on the left in (7.12) gives a measure of the *backward error* in x_k, since if this quantity is bounded by ζ, then x_k is the exact solution of a nearby problem $(A + \delta A)x_k = b$, where $\|\delta A\|/\|A\| \leq \zeta + O(\zeta^2)$. In general, the best one can hope for from a computed "solution" is that its backward error be about the machine precision ϵ, since errors of this order are made in simply representing the matrix A (either through its entries or through a procedure for computing the product of A with a given vector) on the computer.

Based on (7.12), one can expect a small backward error from iterative methods of the form (7.10), if the norms of the iterates x_k do not greatly exceed that of the true solution. For the CG algorithm for Hermitian positive definite linear systems, this is the case. It can be shown in exact arithmetic that the 2-norm of the error decreases monotonically in the CG algorithm [79]. From the inequality $\|x - x_k\| \leq \|x - x_0\|$, it follows that $\|x_k\| \leq 2\|x\| + \|x_0\|$. Assuming that $\|x_0\| \leq \|x\|$, the quantity $\max_k \|x_k\|/\|x\|$ in (7.12) is therefore bounded by about 3, and one can expect to eventually obtain an approximate solution whose backward error is a moderate multiple of ϵ. In finite precision arithmetic, the relation established in Chapter 4 between the error norms in finite precision arithmetic and exact arithmetic error norms for a larger problem can be used to show that a monotone reduction in the 2-norm of the error can be expected (to a close approximation) in finite precision arithmetic as well.

If the BiCG algorithm is implemented in the form (7.10), as in the algorithm stated in section 5.2, the norms of intermediate iterates may grow. Since no special error norm (that can be related easily to the 2-norm) is guaranteed to decrease in the BiCG algorithm, the norms of the iterates cannot be bounded a priori, and growth of the iterates will cause loss of accuracy in the final approximate solution, even assuming that the updated vectors r_k converge to zero.

An example is shown in Figure 7.1. Here A was taken to be a discretization of the convection–diffusion operator

$$- \triangle u + 40(xu_x + yu_y) - 100u$$

on the unit square with Dirichlet boundary conditions, using centered differences on a 32-by-32 mesh. The solution was taken to be $u(x, y) = x(x - 1)^2 y^2 (y - 1)^2$, and the initial guess was set to zero. The initial vector $\hat{r}_0$ was set equal to r_0.

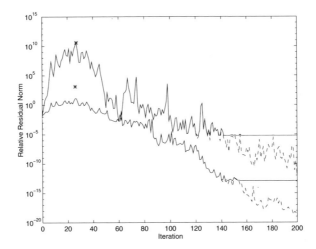

FIG. 7.1. *Actual residual norm (solid) and updated residual norm (dashed). Top curves are for CGS, bottom ones for BiCG. The asterisk shows the maximum ratio* $\|x_k\|/\|x\|$.

The lower solid line in Figure 7.1 represents the true BiCG residual norm $\|b - Ax_k\|/(\|A\|\|x\|)$, while the lower dashed line shows the updated residual norm, $\|r_k\|/(\|A\|\|x\|)$. The lower asterisk in the figure shows the maximum ratio $\|x_k\|/\|x\|$ at the step at which it occurred. The experiment was run on a machine with unit roundoff $\epsilon \approx 1.1e - 16$, and the maximum ratio $\|x_k\|/\|x\|$ was approximately 10^3. As a result, instead of achieving a final residual norm of about ϵ, the final residual norm is about $1.e - 13 \approx 10^3\epsilon$.

Also shown in Figure 7.1 (upper solid and dashed lines) are the results of running the CGS algorithm given in section 5.5 for this same problem. Again, there are no a priori bounds on the size of intermediate iterates, and in this case we had $\max_k \|x_k\|/\|x\| \approx 4 \cdot 10^{10}$. As a result, the final actual residual norm reaches the level $6.e - 6$, which is roughly $4 \cdot 10^{10} \epsilon$.

Since the CGNE and CGNR algorithms are also of the form (7.10), the estimate (7.12) is applicable to them as well. Since CGNE minimizes the 2-norm of the error, it follows, as for CG, that $\|x_k\| \leq 2\|x\| + \|x_0\|$, so the backward error in the final approximation will be a moderate multiple of the machine precision. The CGNR method minimizes the 2-norm of the residual, but since it is equivalent to CG for the linear system $A^H Ax = A^H b$, it follows that the 2-norm of the error also decreases monotonically. Hence we again expect a final backward error of order ϵ.

An example for the CGNE method is shown in Figure 7.2. The matrix A was taken to be of the form $A = U\Sigma V^T$, where U and V are random orthogonal matrices and $\Sigma = \text{diag}(\sigma_1, \ldots, \sigma_n)$, with

$$\sigma_i = \kappa^{(i-1)/(n-1)}, \quad i = 1, \ldots, n, \quad \kappa = 10^4.$$

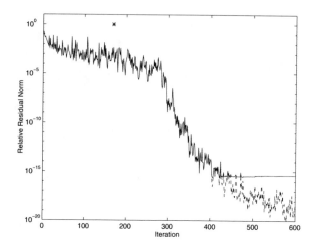

FIG. 7.2. *Actual residual norm (solid) and updated residual norm (dashed) for CGNE. The asterisk shows the maximum ratio* $\|x_k\|/\|x\|$.

For a problem of size $n = 40$, a random solution was set and a zero initial guess was used. The solid line in Figure 7.2 shows the actual residual norm $\|b - Ax_k\|/(\|A\|\|x\|)$, while the dashed line represents the updated residual norm $\|r_k\|/(\|A\|\|x\|)$. The maximum ratio $\|x_k\|/\|x\|$ is approximately 1, as indicated by the asterisk in Figure 7.2. Note that while rounding errors greatly affect the convergence rate of the method—in exact arithmetic, the exact solution would be obtained after 40 steps—the ultimately attainable accuracy is as great as one could reasonably expect—a backward error of size approximately 4ϵ. There is no loss of final accuracy due to the fact that we are (implicitly) solving the normal equations.

When a preconditioner is used with the above algorithms, the 2-norm of the error may not decrease monotonically, and then one must use other properties to establish bounds on the norms of the iterates.

It is sometimes asked whether one can accurately solve a very ill-conditioned linear system if a very good preconditioner is available. That is, suppose $\kappa(A)$ is very large but $\kappa(M^{-1}A)$ or $\kappa(M^{-1/2}AM^{-1/2})$, where M is a known preconditioning matrix, is not. We will see in Part II that the preconditioned CG algorithm, for instance, still uses formulas of the form (7.10). There is simply an additional formula, $Mz_k = r_k$, to determine a preconditioned residual z_k. The final residual norm is still given approximately by (7.12), and, unless the final residual vector is deficient in certain eigencomponents of A, this suggests an error satisfying

$$\min_k \frac{\|x - x_k\|}{\|x\|} \approx c\, \epsilon\, \kappa(A).$$

The presence of even an excellent preconditioner M (such as the LU factors from direct Gaussian elimination) does not appear to improve this error bound for algorithms of the form (7.10).

For a discussion of the effect of rounding errors on the attainable accuracy with some different implementations, see, for example, [33, 70, 122].

7.4. Multiple Right-Hand Sides and Block Methods.

Frequently, it is desired to solve several linear systems with the same coefficient matrix but different right-hand sides. Sometimes the right-hand side vectors are known at the start and sometimes one linear system must be solved before the right-hand side for the next linear system can be computed (as, for example, in time-dependent partial differential equations). One might hope that information gained in the solution of one linear system could be used to facilitate the solution of subsequent problems with the same coefficient matrix.

We will consider only the case in which the right-hand sides are all available at the start. In that case, *block* versions of the previously described algorithms can be used. Suppose there are s right-hand sides. Then the linear systems can be written in the form

$$AX = B,$$

where X is the n-by-s matrix of solution vectors and B is the n-by-s matrix of right-hand sides.

Let A be Hermitian positive definite and consider the *block CG* algorithm. Instead of minimizing the A-norm of the error for each linear system over a single Krylov space, one can minimize

$$\text{tr}[(X - X_k)^H A(X - X_k)]$$

over all X_k of the form

$$X_k \in X_0 + \text{span}[R_0, AR_0, \ldots, A^{k-1}R_0].$$

That is, the approximation $x_k^{(\ell)}$ for the ℓth equation is equal to $x_0^{(\ell)}$ plus a linear combination of vectors from *all* of the Krylov spaces

$$\bigcup_{j=1}^{s} \text{span}\left[r_0^{(j)}, Ar_0^{(j)}, \ldots, A^{k-1}r_0^{(j)}\right]$$

The following algorithm accomplishes this minimization.

Algorithm 8. Block Conjugate Gradient Method (Block CG)
(for Hermitian positive definite problems with multiple right-hand sides).

Given an initial guess X_0, compute $R_0 = B - AX_0$ and set $P_0 = R_0$.
For $k = 1, 2, \ldots$,

Compute AP_{k-1}.

Set $X_k = X_{k-1} + P_{k-1}a_{k-1}$, where $a_{k-1} = (P_{k-1}^H AP_{k-1})^{-1}(R_{k-1}^H R_{k-1})$.

Compute $R_k = R_{k-1} - AP_{k-1}a_{k-1}$.

Set $P_k = R_k + P_{k-1}b_{k-1}$, where $b_{k-1} = (R_{k-1}^H R_{k-1})^{-1}(R_k^H R_k)$.

It is left as an exercise to show that the following block orthogonality properties hold:

$$R_k^H R_j = 0, \quad j \neq k, \quad \text{and} \quad P_k^H AP_j = 0, \quad j \neq k.$$

As long as the matrices P_k and R_k retain full rank, the algorithm is well defined. If their columns become linearly dependent, then equations corresponding to dependent columns can be treated separately, and the algorithm will be continued with the remaining equations. A number of strategies have been developed for varying the block size. See, for example, [107, 130, 105].

A block algorithm requires somewhat more work than running s separate recurrences because s separate inner products and scalar divisions are replaced by the formation of an s-by-s matrix and the solution of s linear systems with this coefficient matrix. (Note, as usual, that it is not necessary to actually invert the matrices $P_{k-1}^H AP_{k-1}$ and $R_{k-1}^H R_{k-1}$ in the block CG algorithm, but instead one can solve s linear systems with right-hand sides given by the columns of $R_{k-1}^H R_{k-1}$ and $R_k^H R_k$, respectively. This can be accomplished by factoring the coefficient matrices using Cholesky decomposition and then backsolving with the triangular factors s times. The total work is $O(s^3)$.) If $s^3 < n$, then the extra work will be negligible.

How much improvement is obtained in the number of iterations required due to the fact that the error is minimized over a larger space? This depends on the right-hand side vectors. If all of the vectors in all of the s Krylov spaces are linearly independent, then at most n/s steps are required for the block algorithm, as compared to n for the nonblock version. Usually, however, the number of iterations required even for the nonblock iteration is significantly less than n/s. In this case, if the right-hand sides are unrelated random vectors, then the improvement in the number of iterations is usually modest. Special relations between right-hand side vectors, however, may lead to significant advantages for the block algorithm.

7.5. Computer Implementation.

If two iterative methods are both capable of generating a sufficiently accurate approximation to a system of linear equations, then we usually compare the two methods by counting operations—how many additions, subtractions, multiplications, and divisions will each method require? If the number of operations per iteration is about the same for the two methods, then we might just compare number of iterations. The method that requires fewer iterations is chosen. (One should be very careful about iteration count comparisons, however, to be sure that the algorithms being compared really do require the same amount of work per iteration!)

These are approximate measures of the relative computer time that will be required by the two algorithms, but they are only approximate. Computational time may also depend on *data locality* and potential for *parallelism*, that is, the ability to effectively use multiple processors simultaneously. These factors vary from one machine to another, so it is generally impossible to give a definitive answer to the question of which algorithm is faster.

Almost all of the algorithms discussed in this book perform vector *inner products* during the course of the iteration. If different pieces of the vectors are stored on different processors, then this requires some *global communication* to add together the inner products of the subvectors computed on the different processors. It has sometimes been thought that this would be a major bottleneck for distributed memory multiprocessors, but on today's supercomputers, this does not appear to be the case. It is a relatively small amount of data that must be passed between processors, and the bulk of the time for the iterative solver still lies in the matrix–vector multiplication and in the preconditioning step.

Sparse matrix–vector multiplication is often parallelized by assigning different rows or different blocks of the matrix to different processors, along with the corresponding pieces of the vectors on which they must operate. Many different distribution schemes are possible.

The most difficult part of an iterative method to parallelize is often the preconditioning step. This may require the solution of a sparse triangular system, which is a largely sequential operation—one solution component must be known before the next can be computed. For this reason, a number of more parallelizable preconditioners have been proposed. Examples include sparse approximate inverses (so the preconditioning step becomes just another sparse matrix–vector multiplication) and domain decomposition methods. Domain decomposition methods, to be discussed in Chapter 12, divide the physical domain of the problem into pieces and assign different pieces to different processors. The preconditioning step involves each processor solving a problem on its subdomain. In order to prevent the number of iterations from growing with the number of subdomains, however, some global communication is required. This is in the form of a coarse grid solve.

A number of parallel iterative method packages have been developed for different machines. Some examples are described in [118, 82]. More information about the parallelization of iterative methods can be found in [117].

Exercises.

7.1. Let A be a matrix of the form $I - F$, where $F = -F^H$. Suppose the eigenvalues of A are contained in the line segment $[1 - i\gamma, 1 + i\gamma]$. It was shown by Freund and Ruscheweyh [55] that if a MINRES algorithm is applied to this matrix, then the residual at step k satisfies

$$(7.13) \qquad \frac{\|r_k\|}{\|r_0\|} \leq \frac{2}{R^k + R^{k-2}}, \qquad R = \frac{1 + \sqrt{1 + \gamma^2}}{\gamma}.$$

Moreover, if A contains eigenvalues throughout the interval, then this bound is sharp.

Determine a bound on the residual in the CGNR method. Will CGNR require more or fewer iterations than the MINRES method for this problem?

7.2. Use the fact that $T_k = Q_k^H A Q_k$ in the Hermitian Lanczos algorithm to show that the eigenvalues of T_k lie between the smallest and largest eigenvalues of A.

7.3. Prove the block orthogonality properties $R_k^H R_j = P_k^H A P_j = 0$, $j \neq k$, for the block CG algorithm.

Part II

Preconditioners

Chapter 8

Overview and Preconditioned Algorithms

All of the iterative methods discussed in Part I of this book converge very rapidly if the coefficient matrix A is close to the identity. Unfortunately, in most applications, A is not close to the identity, but one might consider replacing the original linear system $Ax = b$ by the modified system

$$(8.1) \qquad M^{-1}Ax = M^{-1}b \quad \text{or} \quad AM^{-1}\hat{x} = b, \quad x = M^{-1}\hat{x}.$$

These are referred to as *left* and *right* preconditioning, respectively. If M is Hermitian and positive definite, then one can precondition symmetrically and solve the modified linear system

$$(8.2) \qquad\qquad L^{-1}AL^{-H}y = L^{-1}b, \quad x = L^{-H}y,$$

where $M = LL^H$. The matrix L could be the Hermitian square root of M or the lower triangular Cholesky factor of M or any other matrix satisfying $M = LL^H$. In either case, it is necessary only to be able to solve linear systems with coefficient matrix M, not to actually compute M^{-1} or L.

If the *preconditioner* M can be chosen so that

1. linear systems with coefficient matrix M are easy to solve, and

2. $M^{-1}A$ or AM^{-1} or $L^{-1}AL^{-H}$ approximates the identity,

then an efficient solution technique results from applying an iterative method to the modified linear system (8.1) or (8.2).

The exact sense in which the preconditioned matrix should approximate the identity depends on the iterative method being used. For *simple iteration*, one would like $\rho(I - M^{-1}A) << 1$ to achieve fast asymptotic convergence or $\|I - M^{-1}A\| << 1$ to achieve large error reduction at each step.

For the *CG* or *MINRES* methods for Hermitian positive definite problems, one would like the condition number of the symmetrically preconditioned matrix $L^{-1}AL^{-H}$ to be close to one, in order for the error bound based on the Chebyshev polynomial to be small. Alternatively, a preconditioned matrix with just a few large eigenvalues and the remainder tightly clustered would also be good for the CG and MINRES algorithms, as would a preconditioned matrix with just a few distinct eigenvalues. For MINRES applied to a Hermitian indefinite linear system but with a positive definite preconditioner, it is again the eigenvalue distribution of the preconditioned matrix that is of importance. The eigenvalues should be distributed in such a way that a polynomial of moderate degree with value one at the origin can be made small at all of the eigenvalues.

For *GMRES*, a preconditioned matrix that is close to normal and whose eigenvalues are tightly clustered around some point away from the origin would be good, but other properties might also suffice to define a good preconditioner. It is less clear exactly what properties one should look for in a preconditioner for some of the other non-Hermitian matrix iterations (such as BiCG, QMR, CGS, or BiCGSTAB), but again, since each of these methods converges in one iteration if the coefficient matrix is the identity, there is the intuitive concept that the preconditioned matrix should somehow approximate the identity.

It is easy to modify the algorithms of Part I to use left preconditioning— simply replace A by $M^{-1}A$ and b by $M^{-1}b$ everywhere they appear. Right or symmetric preconditioning requires a little more thought since we want to generate approximations x_k to the solution of the original linear system, not the modified one in (8.1) or (8.2).

If the CG algorithm is applied directly to equation (8.2), then the iterates satisfy

$$y_k = y_{k-1} + a_{k-1}\hat{p}_{k-1}, \quad a_{k-1} = \frac{\langle \hat{r}_{k-1}, \hat{r}_{k-1} \rangle}{\langle \hat{p}_{k-1}, L^{-1}AL^{-H}\hat{p}_{k-1} \rangle},$$

$$\hat{r}_k = \hat{r}_{k-1} - a_{k-1}L^{-1}AL^{-H}\hat{p}_{k-1},$$

$$\hat{p}_k = \hat{r}_k + b_{k-1}\hat{p}_{k-1}, \quad b_{k-1} = \frac{\langle \hat{r}_k, \hat{r}_k \rangle}{\langle \hat{r}_{k-1}, \hat{r}_{k-1} \rangle}.$$

Defining

$$x_k \equiv L^{-H}y_k, \quad r_k \equiv L\hat{r}_k, \quad p_k \equiv L^{-H}\hat{p}_k,$$

we obtain the following preconditioned CG algorithm for $Ax = b$.

Algorithm 2P. Preconditioned Conjugate Gradient Method (PCG) (for Hermitian positive definite problems, with Hermitian positive definite preconditioners).

Given an initial guess x_0, compute $r_0 = b - Ax_0$ and solve $Mz_0 = r_0$. Set $p_0 = z_0$. For $k = 1, 2, \ldots$,

Compute Ap_{k-1}.

Set $x_k = x_{k-1} + a_{k-1}p_{k-1}$, where $a_{k-1} = \frac{\langle r_{k-1}, z_{k-1} \rangle}{\langle p_{k-1}, Ap_{k-1} \rangle}$.

Compute $r_k = r_{k-1} - a_{k-1}Ap_{k-1}$.

Solve $Mz_k = r_k$.

Set $p_k = z_k + b_{k-1}p_{k-1}$, where $b_{k-1} = \frac{\langle r_k, z_k \rangle}{\langle r_{k-1}, z_{k-1} \rangle}$.

The same modifications can be made to any of the MINRES implementations, provided that the preconditioner M is positive definite. To obtain a preconditioned version of Algorithm 4, first consider the Lanczos algorithm applied directly to the matrix $L^{-1}AL^{-H}$ with initial vector $\hat{q}_1$. Successive vectors satisfy

$$\hat{v}_j = L^{-1}AL^{-H}\hat{q}_j - \alpha_j\hat{q}_j - \beta_{j-1}\hat{q}_{j-1},$$

$$\alpha_j = \langle L^{-1}AL^{-H}\hat{q}_j, \hat{q}_j \rangle - \beta_{j-1}\langle \hat{q}_{j-1}, \hat{q}_j \rangle,$$

$$\hat{q}_{j+1} = \hat{v}_j/\beta_j, \quad \beta_j = \|\hat{v}_j\|.$$

If we define $q_j \equiv L\hat{q}_j$, $v_j \equiv L\hat{v}_j$, and $w_j \equiv M^{-1}q_j$, then the same equations can be written in terms of q_j, v_j, and w_j.

Preconditioned Lanczos Algorithm (for Hermitian matrices A, with Hermitian positive definite preconditioners M).

Given v_0, solve $M\tilde{w}_1 = v_0$, and set $\beta_0 = \langle v_0, \tilde{w}_1 \rangle^{1/2}$.
Set $q_1 = v_0/\beta_0$ and $w_1 = \tilde{w}_1/\beta_0$. Define $q_0 \equiv 0$. For $j = 1, 2, \ldots$,

Set $v_j = Aw_j - \beta_{j-1}q_{j-1}$.

Compute $\alpha_j = \langle v_j, w_j \rangle$, and update $v_j \leftarrow v_j - \alpha_j q_j$.

Solve $M\tilde{w}_{j+1} = v_j$.

Set $q_{j+1} = v_j/\beta_j$ and $w_{j+1} = \tilde{w}_{j+1}/\beta_j$, where $\beta_j = \langle v_j, \tilde{w}_{j+1} \rangle^{1/2}$.

If Algorithm 4 of section 2.5 is applied directly to the preconditioned linear system (8.2) and if we let y_k and $\hat{p}_k$ denote the iterates and direction vectors generated by that algorithm and if we then define $x_k \equiv L^{-H} y_k$ and $p_k \equiv L^{-H} \hat{p}_k$, then these vectors are generated by the following preconditioned algorithm.

Algorithm 4P. Preconditioned Minimal Residual Algorithm (PMINRES)
(for Hermitian problems, with Hermitian positive definite preconditioners).

Given x_0, compute $r_0 = b - A x_0$ and solve $M z_0 = r_0$.
Set $\beta = \langle r_0, z_0 \rangle^{1/2}$, $q_1 = r_0/\beta$, and $w_1 = z_0/\beta$.
Initialize $\xi = (1, 0, \ldots, 0)^T$. For $k = 1, 2, \ldots,$

Compute q_{k+1}, w_{k+1}, $\alpha_k \equiv T(k, k)$, and $\beta_k \equiv T(k+1, k) \equiv T(k, k+1)$ using the preconditioned Lanczos algorithm.

Apply F_{k-2} and F_{k-1} to the last column of T; that is

$$\begin{pmatrix} T(k-2, k) \\ T(k-1, k) \end{pmatrix} \leftarrow \begin{pmatrix} c_{k-2} & s_{k-2} \\ -\bar{s}_{k-2} & c_{k-2} \end{pmatrix} \begin{pmatrix} 0 \\ T(k-1, k) \end{pmatrix}, \quad \text{if } k > 2,$$

$$\begin{pmatrix} T(k-1, k) \\ T(k, k) \end{pmatrix} \leftarrow \begin{pmatrix} c_{k-1} & s_{k-1} \\ -\bar{s}_{k-1} & c_{k-1} \end{pmatrix} \begin{pmatrix} T(k-1, k) \\ T(k, k) \end{pmatrix}, \quad \text{if } k > 1.$$

Compute the kth rotation, c_k and s_k, to annihilate the $(k+1, k)$ entry of T.[1]

Apply kth rotation to ξ and to last column of T:

$$\begin{pmatrix} \xi(k) \\ \xi(k+1) \end{pmatrix} \leftarrow \begin{pmatrix} c_k & s_k \\ -\bar{s}_k & c_k \end{pmatrix} \begin{pmatrix} \xi(k) \\ 0 \end{pmatrix}.$$

$$T(k, k) \leftarrow c_k T(k, k) + s_k T(k+1, k), \quad T(k+1, k) \leftarrow 0.$$

Compute $p_{k-1} = [w_k - T(k-1, k) p_{k-2} - T(k-2, k) p_{k-3}]/T(k, k)$, where undefined terms are zero for $k \leq 2$.

Set $x_k = x_{k-1} + a_{k-1} p_{k-1}$, where $a_{k-1} = \beta \xi(k)$.

Right-preconditioned algorithms for non-Hermitian matrices are similarly derived so that the algorithms actually generate and store approximations to the solution of the original linear system.

Preconditioners can be divided roughly into three categories:

I. Preconditioners designed for general classes of matrices; e.g., matrices with nonzero diagonal entries, positive definite matrices, M-matrices. Examples of such preconditioners are the Jacobi, Gauss–Seidel, and

---

[1]The formula is $c_k = |T(k, k)|/\sqrt{|T(k, k)|^2 + |T(k+1, k)|^2}$, $\bar{s}_k = c_k T(k+1, k)/T(k, k)$, but a more robust implementation should be used. See, for example, BLAS routine DROTG [32].

SOR preconditioners, the incomplete Cholesky, and modified incomplete Cholesky preconditioners.

II. Preconditioners designed for broad classes of underlying problems; e.g., elliptic partial differential equations. Examples are multigrid and domain decomposition preconditioners.

III. Preconditioners designed for a specific matrix or underlying problem; e.g., the transport equation. An example is the diffusion synthetic acceleration (DSA) preconditioner, which will be mentioned but not analyzed in section 9.2.

An advantage of category I preconditioners is that they can be used in settings where the exact origin of the problem is not necessarily known— for example, in software packages for solving systems of ordinary differential equations or optimization problems. Most of the preconditioners in category I require knowledge of at least some of the entries of A. Usually, this information is readily available, but sometimes it is much easier to compute matrix–vector products, through some special formula, than it is to compute the actual entries of the matrix (in the standard basis). For example, one might use a finite difference approximation to the product of a Jacobian matrix with a given vector, without ever computing the Jacobian itself [23]. For such problems, efficient general preconditioners may be difficult to derive, and we do not address this topic here.

For practical preconditioners designed for general matrices, there are few quantitative theorems to describe just how good the preconditioner is, e.g., how much smaller the condition number of the preconditioned matrix is compared to that of the original matrix. There are, however, comparison theorems available. For large classes of problems, one may be able to prove that a preconditioner M_1 is better than another preconditioner M_2, in that, say, $\rho(I - M_1^{-1}A) < \rho(I - M_2^{-1}A)$. Such results are discussed in Chapter 10. These results may lead to theorems about the optimal preconditioner of a given form; e.g., the optimal diagonal preconditioner.

For classes of problems arising from partial differential equations, it is sometimes possible to show that a preconditioner alters the dependence of the condition number on the mesh size used in a finite difference or finite element approximation. That is, instead of considering a single matrix A and asking how much a particular preconditioner reduces the condition number of A, we consider a class of matrices A_h and preconditioners M_h parameterized by a mesh spacing h. It can sometimes be shown that while the condition number of A_h grows like $O(h^{-2})$ as $h \to 0$, the condition number of $M_h^{-1/2} A_h M_h^{-1/2}$ is only $O(h)$ or $O(1)$. This is not of much help if one's goal is to solve a specific linear system $Ax = b$, but if the goal is to solve the underlying partial differential equation, it quantifies the difficulty of solving the linear system in relation to the accuracy of the finite difference or finite element scheme. In Chapter 11, incomplete decompositions are considered, and it is shown that for a model problem, a modified incomplete Cholesky decomposition reduces the condition number of A from $O(h^{-2})$ to $O(h^{-1})$. Multigrid methods, discussed in Chapter 12, have proved especially effective for solving problems arising from partial differential equations because they often eliminate the dependence of the condition number on h entirely.

Despite great strides in developing preconditioners for general linear systems or for broad classes of underlying problems, it is still possible in many situations to use physical intuition about a specific problem to develop a more effective preconditioner. Note that this is different from the situation with the iterative techniques themselves. Seldom (if ever) can one use physical properties of the problem being solved to devise an iteration strategy (i.e., a choice of the polynomial P_k for which $r_k = P_k(A)r_0$) that is better than, say, the CG method. For this reason, the subject of preconditioners is still a very broad one, encompassing all areas of science. No complete survey can be given. In this book, we present some known general theory about preconditioners and a few example problems to illustrate their use in practice.

Comments and Additional References.

The idea of preconditioning the CG method actually appeared in the original Hestenes and Stiefel paper [79]. It was not widely used until much later, however, after works such as [27] and [99]. See also [87], which describes some early applications. It was the development of effective preconditioning strategies that helped bring the CG algorithm into widespread use as an iterative method.

Chapter 9

Two Example Problems

In the study of preconditioners, it is useful to have some specific problems in mind. Here we describe two such problems—one of which (the diffusion equation) gives rise to a symmetric positive definite linear system, and one of which (the transport equation) gives rise to a nonsymmetric linear system. Many other examples could equally well have been chosen for presentation, but these two problems are both physically important and illustrative of many of the principles to be discussed. Throughout this chapter we will deal only with *real* matrices.

9.1. The Diffusion Equation.

A number of different physical processes can be described by *the diffusion equation*:

$$(9.1) \qquad \frac{\partial u}{\partial t} - \nabla \cdot (a \nabla u) = f \quad \text{in } \Omega.$$

Here u might represent the temperature distribution at time t in an object Ω, to which an external heat source f is applied. The positive coefficient $a(\mathbf{x})$ is the thermal conductivity of the material. To determine the temperature at time t, we need to know an initial temperature distribution $u(\mathbf{x}, 0)$ and some boundary conditions, say,

$$(9.2) \qquad u(\mathbf{x}, t) = 0 \quad \text{on } \partial\Omega,$$

corresponding to the boundary of the region being held at a fixed temperature (which we have denoted as 0).

Other phenomena lead to an equation of the same form. For example, equation (9.1) also represents the diffusion of a substance through a permeable region Ω, if u is interpreted as the concentration of the substance, a as the diffusion coefficient of the material, and f as the specific rate of generation of the substance by chemical reactions or outside sources.

A standard method for obtaining approximate solutions to partial differential equations such as (9.1) is the method of *finite differences*. Here the region Ω is divided into small pieces, and at each point of a grid on Ω, the derivatives in (9.1) are replaced by difference quotients that approach the true derivatives as the grid becomes finer.

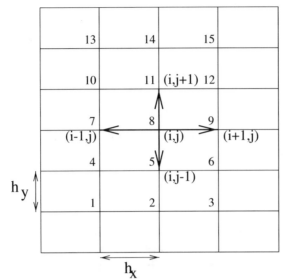

Fig. 9.1. *Finite difference discretization, natural ordering.*

For example, suppose the region Ω is the unit square $[0, 1] \times [0, 1]$. Introduce a uniform grid $\{x_i, y_j : i = 0, 1, \ldots, n_x + 1, j = 0, 1, \ldots, n_y + 1\}$ with spacing $h_x = 1/(n_x + 1)$ in the x-direction and $h_y = 1/(n_y + 1)$ in the y-direction, as shown in Figure 9.1 for $n_x = 3$, $n_y = 5$. A standard centered difference approximation to the partial derivative in the x-direction in (9.1) is

$$\left(\frac{\partial}{\partial x} a \frac{\partial u}{\partial x} \right)(x_i, y_j) \approx \frac{a_{i+1/2,j}(u_{i+1,j} - u_{i,j}) - a_{i-1/2,j}(u_{i,j} - u_{i-1,j})}{h_x^2},$$

where $a_{i\pm1/2,j} \equiv a(x_i \pm h_x/2, y_j)$ and $u_{i,j}$ represents the approximation to $u(x_i, y_j)$. An analogous expression is obtained for the partial derivative in the y direction:

$$\left(\frac{\partial}{\partial y} a \frac{\partial u}{\partial y} \right)(x_i, y_j) \approx \frac{a_{i,j+1/2}(u_{i,j+1} - u_{i,j}) - a_{i,j-1/2}(u_{i,j} - u_{i,j-1})}{h_y^2},$$

where $a_{i,j\pm1/2} \equiv a(x_i, y_j \pm h_y/2)$. We will sometimes be interested in problems for which $a(x, y)$ is *discontinuous* along a mesh line. In such cases, the values $a_{i\pm1/2,j}$ and $a_{i,j\pm1/2}$ will be taken to be averages of the surrounding values. For instance, if $a(x, y)$ is discontinuous along the line $y = y_j$, then $a_{i\pm1/2,j} \equiv \lim_{\epsilon \to 0+}(a(x_i \pm 1/2h_x, y_j + \epsilon) + a(x_i \pm 1/2h_x, y_j - \epsilon)/2$.

If the steady-state version of problem (9.1–9.2),

$$-\nabla \cdot (a \nabla u) = f \quad \text{in } \Omega \equiv (0, 1) \times (0, 1),$$

$$u(x, 0) = u(x, 1) = u(0, y) = u(1, y) = 0,$$

is approximated by this finite difference technique, then we obtain the following system of $n_x n_y$ linear algebraic equations to solve for the unknown function values $u_{i,j}$ at the interior mesh points:

$$-\left(\frac{a_{i+1/2,j}(u_{i+1,j} - u_{i,j}) - a_{i-1/2,j}(u_{i,j} - u_{i-1,j})}{h_x^2} +\right.$$

$$\left.\frac{a_{i,j+1/2}(u_{i,j+1} - u_{i,j}) - a_{i,j-1/2}(u_{i,j} - u_{i,j-1})}{h_y^2}\right) = f_{i,j},$$

(9.3) $i = 1, \ldots, n_x, \quad j = 1, \ldots, n_y.$

For the time-dependent equation, a backward or centered difference approximation in time is often used, resulting in a system of linear algebraic equations to solve at each time step. For example, if the solution $u_{i,j}^\ell$ at time t_ℓ is known, and if backward differences in time are used, then in order to obtain the approximate solution $u_{i,j}^{\ell+1}$ at time $t_{\ell+1} = t_\ell + \Delta t$, one must solve the following system of equations:

$$\frac{u_{i,j}^{\ell+1} - u_{i,j}^\ell}{\Delta t} - \left(\frac{a_{i+1/2,j}\left(u_{i+1,j}^{\ell+1} - u_{i,j}^{\ell+1}\right) - a_{i-1/2,j}\left(u_{i,j}^{\ell+1} - u_{i-1,j}^{\ell+1}\right)}{h_x^2} +\right.$$

$$\left.\frac{a_{i,j+1/2}\left(u_{i,j+1}^{\ell+1} - u_{i,j}^{\ell+1}\right) - a_{i,j-1/2}\left(u_{i,j}^{\ell+1} - u_{i,j-1}^{\ell+1}\right)}{h_y^2}\right) = f_{i,j}^{\ell+1},$$

(9.4) $i = 1, \ldots, n_x, \quad j = 1, \ldots, n_y.$

Here we have considered a two-dimensional problem for illustration, but it should be noted that iterative methods are especially important for three-dimensional problems, where direct methods become truly prohibitive in terms of both time and storage. The extension of the difference scheme to the unit cube is straightforward.

To write the equations (9.3) or (9.4) in matrix form, we must choose an ordering for the equations and unknowns. A common choice, known as the *natural ordering*, is to number the gridpoints from left to right and bottom to top, as shown in Figure 9.1. With this ordering, equations (9.3) can be written in the form

(9.5) $A\mathbf{u} = \mathbf{f},$

where A is a block tridiagonal matrix with n_y diagonal blocks, each of dimension n_x by n_x; $\mathbf{u}$ is the $n_x n_y$-vector of function values with $u_{i,j}$ stored in position $(j-1)n_x + i$; and $\mathbf{f}$ is the $n_x n_y$-vector of right-hand side values with $f_{i,j}$ in position $(j-1)n_x + i$. Define

$$d_{i,j} \equiv \frac{a_{i+1/2,j} + a_{i-1/2,j}}{h_x^2} + \frac{a_{i,j+1/2} + a_{i,j-1/2}}{h_y^2},$$

(9.6) $$b_{i+1/2,j} \equiv \frac{-a_{i+1/2,j}}{h_x^2}, \quad c_{i,j+1/2} \equiv \frac{-a_{i,j+1/2}}{h_y^2}.$$

Then the coefficient matrix A can be written in the form

(9.7) $$A = \begin{pmatrix} S_1 & T_{3/2} & & & \\ T_{3/2} & \ddots & & \ddots & \\ & \ddots & & \ddots & T_{n_y-1/2} \\ & & T_{n_y-1/2} & S_{n_y} \end{pmatrix},$$

where

$$S_j = \begin{pmatrix} d_{1,j} & b_{3/2,j} & & \\ b_{3/2,j} & \ddots & & \ddots \\ & \ddots & & \ddots & b_{n_x-1/2,j} \\ & & b_{n_x-1/2,j} & d_{n_x,j} \end{pmatrix},$$

(9.8) $$T_{j+1/2} = \begin{pmatrix} c_{1,j+1/2} & & & \\ & \ddots & & \\ & & \ddots & \\ & & & c_{n_x,j+1/2} \end{pmatrix}.$$

For the time-dependent problem (9.4), the diagonal entries of A are increased by $1/\Delta t$, and the terms $u_{i,j}^\ell/\Delta t$ are added to the right-hand side vector.

THEOREM 9.1.1. *Assume that $a(x,y) \geq \alpha > 0$ in $(0,1) \times (0,1)$. Then the coefficient matrix A defined in (9.5–9.8) is symmetric and positive definite.*

Proof. Symmetry is obvious. The matrix is weakly diagonally dominant, so by Gerschgorin's theorem (Theorem 1.3.11) its eigenvalues are all greater than or equal to zero. Suppose there is a nonzero vector $\mathbf{v}$ such that $A\mathbf{v} = 0$, and suppose that the component of $\mathbf{v}$ with the largest absolute value is the one corresponding to the (i,j) grid point. We can choose the sign of $\mathbf{v}$ so that this component is positive. From the definition of A and the assumption that $a(x,y) > 0$, it follows that $v_{i,j}$ can be written as a weighted average of the surrounding values of v:

$$v_{i,j} = w_{i-1,j}v_{i-1,j} + w_{i+1,j}v_{i+1,j} + w_{i,j-1}v_{i,j-1} + w_{i,j+1}v_{i,j+1},$$

$$w_{i\pm1,j} \equiv \frac{1}{d_{i,j}}\frac{a_{i\pm1/2,j}}{h_x^2}, \quad w_{i,j\pm1} \equiv \frac{1}{d_{i,j}}\frac{a_{i,j\pm1/2}}{h_y^2},$$

where terms corresponding to boundary nodes are replaced by zero. The weights $w_{i\pm1,j}$ and $w_{i,j\pm1}$ are positive and sum to 1. It follows that if all neighbors of $v_{i,j}$ are interior points, then they must all have the same maximum value since none can be greater than $v_{i,j}$. Repeating this argument for neighboring points, we eventually find a point with this same maximum value which has at least one neighbor on the boundary. But now the value of $\mathbf{v}$ at this point is a weighted sum of neighboring interior values, where the sum of the weights is less than 1. It follows that the value of $\mathbf{v}$ at one of these other interior points must be greater than $v_{i,j}$ if $v_{i,j} > 0$, which is a contradiction. Therefore the only vector $\mathbf{v}$ for which $A\mathbf{v} = 0$ is the zero vector, and A is positive definite.   □

It is clear that the coefficient matrix for the time dependent problem (9.4) is also positive definite, since it is *strictly* diagonally dominant.

The argument used in Theorem 9.1.1 is a type of *discrete maximum principle*. Note that it did not make use of the specific values of the entries of A—only that A has positive diagonal entries and nonpositive off-diagonal entries (so that the weights in the weighted average are positive); that A is rowwise weakly diagonally dominant, with strong diagonal dominance in at least one row; and that starting from any point (i, j) in the grid, one can reach any other point through a path connecting nearest neighbors. This last property will be associated with an *irreducible* matrix to be defined in section 10.2.

Other orderings of the equations and unknowns are also possible. These change the appearance of the matrix but, provided that the equations and unknowns are ordered in the same way—that is, provided that the rows and columns of A are permuted symmetrically to form a matrix $P^T A P$— the eigenvalues remain the same. For example, if the nodes of the grid in Figure 9.1 are colored in a checkerboard fashion, with red nodes coupling only to black nodes and vice versa, then if the red nodes are ordered first and the black nodes second, then the matrix A takes the form

$$(9.9) \qquad\qquad A = \begin{pmatrix} D_1 & B \\ B^T & D_2 \end{pmatrix},$$

where D_1 and D_2 are diagonal matrices.

A matrix of the form (9.6–9.8) is sometimes called a 5-*point* approximation, since the second derivatives at a point (i, j) are approximated in terms of the function values at that point and its four neighbors. A more accurate approximation can be obtained with a 9-point approximation, coupling function values at each point with its eight nearest neighbors. Another approach to obtaining approximate solutions to partial differential equations is the *finite element method*. The idea of a finite element method is to approximate the solution by a piecewise polynomial—piecewise linear functions on triangles or piecewise bilinear functions on rectangles, etc.—and then to choose the piecewise polynomial to minimize a certain error norm (usually the A-norm of the difference between the true and approximate solution). For piecewise constant $a(x, y)$, the 5-point finite difference matrix turns out to be the same as the matrix arising from a piecewise linear finite element approximation.

9.1.1. Poisson's Equation.   In the special case when the diffusion coefficient $a(x, y)$ is *constant*, say, $a(x, y) \equiv 1$, the coefficient matrix (with the natural ordering of nodes) for the steady-state problem (now known as *Pois-*

son's equation) takes on a very special form:

$$(9.10) \qquad A = \begin{pmatrix} S & T & & \\ T & \ddots & \ddots & \\ & \ddots & \ddots & T \\ & & T & S \end{pmatrix},$$

where $T = (-1/h_y^2)I$ and

$$(9.11) \qquad S = \begin{pmatrix} d & b & & \\ b & \ddots & \ddots & \\ & \ddots & \ddots & b \\ & & b & d \end{pmatrix}, \quad d = \frac{2}{h_x^2} + \frac{2}{h_y^2}, \quad b = \frac{-1}{h_x^2}.$$

This is known as a *block-TST* matrix, where "TST" stands for Toeplitz (constant along diagonals), symmetric, tridiagonal [83]. It is a *block-*TST matrix because the blocks along a given diagonal of the matrix are the same, the matrix is symmetric and block tridiagonal, and each of the blocks is a TST matrix. The eigenvalues and eigenvectors of such matrices are known explicitly.

LEMMA 9.1.1. *Let G be an m-by-m TST matrix with diagonal entries α and off-diagonal entries β. Then the eigenvalues of G are*

$$(9.12) \qquad \lambda_k = \alpha + 2\beta \cos\left(\frac{k\pi}{m+1}\right), \quad k = 1, \ldots, m,$$

and the corresponding orthonormal eigenvectors are

$$(9.13) \qquad q_\ell^{(k)} = \sqrt{\frac{2}{m+1}} \sin\left(\frac{\ell k\pi}{m+1}\right), \quad \ell, k = 1, \ldots, m.$$

Proof. It is easy to verify (9.12–9.13) from the definition of a TST matrix, but here we provide a derivation of these formulas.

Assume that $\beta \neq 0$, since otherwise G is just a multiple of the identity and the lemma is trivial. Suppose λ is an eigenvalue of G with corresponding eigenvector q. Letting $q_0 = q_{m+1} = 0$, we can write $Aq = \lambda q$ in the form

$$(9.14) \qquad \beta q_{\ell-1} + (\alpha - \lambda)q_\ell + \beta q_{\ell+1} = 0, \quad \ell = 1, \ldots, m.$$

This is a linear difference equation, and it can be solved similarly to a corresponding linear differential equation. Specifically, we consider the characteristic polynomial

$$\chi(z) \equiv \beta + (\alpha - \lambda)z + \beta z^2.$$

If the roots of this polynomial are denoted z_+ and z_-, then the general solution of the difference equation (9.14) can be seen to be

$$q_\ell = c_1 z_+^\ell + c_2 z_-^\ell, \quad c_1, c_2 \text{ constants,}$$

and the constants are determined by the boundary conditions $q_0 = q_{m+1} = 0$.
The roots of $\chi(z)$ are

$$(9.15) \qquad z_{\pm} = \frac{\lambda - \alpha \pm \sqrt{(\lambda - \alpha)^2 - 4\beta^2}}{2\beta},$$

and the condition $q_0 = 0$ implies $c_1 + c_2 = 0$. The condition $q_{m+1} = 0$ implies $z_+^{m+1} = z_-^{m+1}$. There are $m + 1$ solutions to this equation, namely,

$$(9.16) \qquad z_+ = z_- \exp\left(\frac{2\pi k\imath}{m+1}\right), \quad k = 0, 1, \ldots, m, \quad \imath \equiv \sqrt{-1},$$

but the $k = 0$ case can be discarded because it corresponds to $z_+ = z_-$ and hence $q_\ell \equiv 0$.

Multiplying by $\exp(-\pi k\imath/(m+1))$ in (9.16) and substituting the values of $z_{\pm}$ from (9.15) yields

$$\left(\lambda - \alpha + \sqrt{(\lambda - \alpha)^2 - 4\beta^2}\right) \exp\left(\frac{-\pi k\imath}{m+1}\right) =$$

$$\left(\lambda - \alpha - \sqrt{(\lambda - \alpha)^2 - 4\beta^2}\right) \exp\left(\frac{\pi k\imath}{m+1}\right).$$

Rearranging, we find

$$\sqrt{(\lambda - \alpha)^2 - 4\beta^2} \, \cos\left(\frac{k\pi}{m+1}\right) = (\lambda - \alpha) \, \imath \, \sin\left(\frac{k\pi}{m+1}\right),$$

and squaring both sides and solving the quadratic equation for λ gives

$$\lambda = \alpha \pm 2\beta \cos\left(\frac{\pi k}{m+1}\right).$$

Taking the plus sign we obtain (9.12), while the minus sign repeats these same values and can be discarded.

Substituting (9.12) for λ in (9.15), we find

$$z_{\pm} = \cos\left(\frac{k\pi}{m+1}\right) \pm \imath \sin\left(\frac{k\pi}{m+1}\right),$$

and therefore

$$q_\ell^{(k)} = c_1(z_+^\ell - z_-^\ell) = 2c_1\imath \sin\left(\frac{\pi k\ell}{m+1}\right), \quad k, \ell = 1, \ldots, m.$$

If we take $c_1 = -(\imath/2)\sqrt{2/(m+1)}$, as in (9.13), then it is easy to check that each vector $q^{(k)}$ has norm one. The eigenvectors are orthogonal since the matrix is symmetric. $\quad\square$

COROLLARY 9.1.1. *All m-by-m TST matrices commute with each other.*

Proof. According to (9.13), all such matrices have the same orthonormal eigenvectors. If $G_1 = Q\Lambda_1 Q^T$ and $G_2 = Q\Lambda_2 Q^T$, then $G_1 G_2 = Q\Lambda_1\Lambda_2 Q^T = Q\Lambda_2\Lambda_1 Q^T = G_2 G_1$. $\square$

THEOREM 9.1.2. *The eigenvalues of the matrix A defined in (9.10–9.11) are*

$$\lambda_{j,k} = \frac{4}{h_x^2} \sin^2\left(\frac{j\pi}{2(n_x+1)}\right) + \frac{4}{h_y^2}\sin^2\left(\frac{k\pi}{2(n_y+1)}\right),$$

(9.17) $$j = 1,\ldots,n_x, \quad k = 1,\ldots,n_y,$$

and the corresponding eigenvectors are

$$u_{m,\ell}^{(j,k)} = \frac{2}{\sqrt{(n_x+1)(n_y+1)}}\sin\left(\frac{mj\pi}{n_x+1}\right)\sin\left(\frac{\ell k\pi}{n_y+1}\right),$$

(9.18) $$m, j = 1,\ldots,n_x, \quad \ell, k = 1,\ldots,n_y,$$

where $u_{m,\ell}^{(j,k)}$ denotes the component corresponding to grid point (m,ℓ) in the eigenvector associated with $\lambda_{j,k}$.

Proof. Let λ be an eigenvalue of A with corresponding eigenvector u, which can be partitioned in the form

$$u \equiv \begin{pmatrix} u_1 \\ \vdots \\ u_{n_y} \end{pmatrix}, \quad u_\ell \equiv \begin{pmatrix} u_{1,\ell} \\ \vdots \\ u_{n_x,\ell} \end{pmatrix}, \quad \ell = 1,\ldots,n_y.$$

The equation $Au = \lambda u$ can be written in the form

(9.19) $$Tu_{\ell-1} + (S - \lambda I)u_\ell + Tu_{\ell+1} = 0, \quad \ell = 1,\ldots,n_y,$$

where we have set $u_0 = u_{n_y+1} = 0$. From Lemma 9.1.1, we can write $S = Q\Lambda_S Q^T$ and $T = Q\Lambda_T Q^T$, where Λ_S and Λ_T are diagonal, with jth-diagonal entries

$$\Lambda_{S,j} = \frac{2}{h_x^2} + \frac{2}{h_y^2} - \frac{2}{h_x^2}\cos\left(\frac{j\pi}{n_x+1}\right), \quad \Lambda_{T,j} = \frac{-1}{h_y^2}.$$

The mth entry of column j of Q is

$$q_m^{(j)} = \sqrt{\frac{2}{n_x+1}}\sin\left(\frac{mj\pi}{n_x+1}\right), \quad m, j = 1,\ldots n_x.$$

Multiply (9.19) by Q^T on the left to obtain

$$\Lambda_T y_{\ell-1} + (\Lambda_S - \lambda I)y_\ell + \Lambda_T y_{\ell+1} = 0, \quad y_\ell \equiv Q^T u_\ell, \quad \ell = 1,\ldots,n_y.$$

Since the matrices here are diagonal, equations along different *vertical* lines in the grid decouple:

(9.20) $$\Lambda_{T,j}y_{j,\ell+1} + \Lambda_{S,j}y_{j,\ell} + \Lambda_{T,j}y_{j,\ell-1} = \lambda y_{j,\ell}, \quad j = 1,\ldots,n_x.$$

If, for a fixed value of j, the vector $(y_{j,1}, \ldots y_{j,n_y})^T$ is an eigenvector of the TST matrix

$$\begin{pmatrix} \lambda_{S,j} & \lambda_{T,j} & & \\ \lambda_{T,j} & \ddots & \ddots & \\ & \ddots & \ddots & \lambda_{T,j} \\ & & \lambda_{T,j} & \lambda_{S,j} \end{pmatrix},$$

with corresponding eigenvalue λ, and if the other components of the vector y are 0, then equations (9.20) will be satisfied. By Lemma 9.1.1, the eigenvalues of this matrix are

$$\begin{aligned}
\lambda_{j,k} &= \lambda_{S,j} + 2\lambda_{T,j} \cos\left(\frac{k\pi}{n_y + 1}\right) \\
&= \frac{2}{h_x^2} + \frac{2}{h_y^2} - \frac{2}{h_x^2}\cos\left(\frac{j\pi}{n_x + 1}\right) - \frac{2}{h_y^2}\cos\left(\frac{k\pi}{n_y + 1}\right) \\
&= \frac{4}{h_x^2}\sin^2\left(\frac{j\pi}{2(n_x + 1)}\right) + \frac{4}{h_y^2}\sin^2\left(\frac{k\pi}{2(n_y + 1)}\right).
\end{aligned}$$

The corresponding eigenvectors are

$$y_{j,\ell}^{(j,k)} = \sqrt{\frac{2}{n_y + 1}}\sin\left(\frac{\ell k\pi}{n_y + 1}\right).$$

Since the ℓth block of $u^{(j,k)}$ is equal to Q times the ℓth block of y and since only the jth entry of the ℓth block of y is nonzero, we have

$$u_{m,\ell}^{(j,k)} = q_m^{(j)} y_{j,\ell}^{(j,k)} = \frac{2}{\sqrt{(n_x + 1)(n_y + 1)}}\sin\left(\frac{mj\pi}{n_x + 1}\right)\sin\left(\frac{\ell k\pi}{n_y + 1}\right).$$

Deriving the eigenvalues $\lambda_{j,k}$ and corresponding vectors $u^{(j,k)}$ for each $j = 1, \ldots, n_x$, we obtain all $n_x n_y$ eigenpairs of A. $\quad\square$

COROLLARY 9.1.2. *Assume that $h_x = h_y \equiv h$. Then the smallest and largest eigenvalues of A in (9.10–9.11) behave like*

(9.21) $$2\pi^2 + O(h^2) \quad and \quad 8h^{-2} + O(1)$$

as $h \to 0$, so the condition number of A is $(4/\pi^2)h^{-2} + O(1)$.

Proof. The smallest eigenvalue of A is the one with $j = k = 1$ and the largest is the one with $j = k = n_x = n_y$ in (9.17):

$$\lambda_{min} = 8h^{-2}\sin^2\left(\frac{\pi h}{2}\right), \quad \lambda_{max} = 8h^{-2}\sin^2\left(\frac{\pi}{2} - \frac{\pi h}{2}\right).$$

Expanding $\sin(x)$ and $\sin(\pi/2 - x)$ in a Taylor series gives the desired result (9.21), and dividing λ_{max} by λ_{min} gives the condition number estimate. $\quad\square$

The proof of Theorem 9.1.4 provides the basis for a direct solution technique for Poisson's equation known as a *fast Poisson solver*. The idea is to separate the problem into individual tridiagonal systems that can be solved independently. The only difficult part is then applying the eigenvector matrix Q to the vectors y obtained from the tridiagonal systems, and this is accomplished using the *fast Fourier transform*. We will not discuss fast Poisson solvers here but refer the reader to [83] for a discussion of this subject.

Because the eigenvalues and eigenvectors of the 5-point finite difference matrix for Poisson's equation on a square are known, preconditioners are often analyzed and even tested numerically on this particular problem, known as the *model problem*. It should be noted, however, that *except for multigrid methods, none of the preconditioned iterative methods discussed in this book is competitive with a fast Poisson solver for the model problem.* The advantage of iterative methods is that they can be applied to more general problems, such as the diffusion equation with a nonconstant diffusion coefficient, Poisson's equation on an irregular region, or Poisson's equation with a nonuniform grid. Fast Poisson solvers apply only to block-TST matrices. They are sometimes used as preconditioners in iterative methods for solving more general problems. Analysis of a preconditioner for the model problem is useful, only to the extent that it can be expected to carry over to more general situations.

9.2. The Transport Equation.

The transport equation is an integro-differential equation that describes the motion of particles (neutrons, photons, etc.) that move in straight lines with constant speed between collisions but which are subject to a certain probability of colliding with outside objects and being scattered, slowed down, absorbed, or multiplied. A sufficiently large aggregate of particles is treated so that they may be regarded as a continuum, and statistical fluctuations are ignored. In the most general setting, the unknown neutron flux is a function of spatial position $r \equiv (x, y, z)$, direction $\Omega \equiv (\sin\theta\cos\phi, \sin\theta\sin\phi, \cos\theta)$, energy E, and time t. Because of the large number of independent variables, the transport equation is seldom solved numerically in its most general form. Instead, a number of approximations are made.

First, a finite number of energy groups are considered and integrals over energy are replaced by sums over the groups. This results in a weakly coupled set of equations for the flux associated with each energy group. These equations are usually solved by a method that we will later identify as a *block Gauss–Seidel* iteration. The flux in the highest energy group is calculated using previously computed approximations for the other energy groups. This newly computed flux is then substituted into the equation for the next energy group, and so on, down to the lowest energy group, at which point the entire process is repeated until convergence. We will be concerned with the mono-energetic transport equation that must be solved for each energy group, at each step of this outer iteration.

The mono-energetic transport equation with *isotropic scattering* can be written as

$$\frac{1}{v}\frac{\partial\psi}{\partial t} + \Omega\cdot\nabla\psi + \sigma_t\psi - \sigma_s\phi = f(r,\Omega,t),$$

(9.22)
$$\phi \equiv \frac{1}{4\pi}\int_{S^2}\psi(r,\Omega',t)\,d\Omega'.$$

Here ψ is the unknown angular flux corresponding to a fixed speed v, σ_t is the known total cross section, σ_s is the known scattering cross section of the material, and f is a known external source. The scalar flux ϕ is the angular flux integrated over directions on the unit sphere S^2. (Actually, the scalar flux is defined without the factor $1/(4\pi)$ in (9.22), but we will include this factor for convenience.) Initial values $\psi(r,\Omega,0)$ and boundary values are needed to specify the solution. If the problem is defined on a region $\mathcal{R}$ with outward normal $n(r)$ at point r, then the incoming flux can be specified by

(9.23) $$\psi(r,\Omega,t) = \psi_g(r,\Omega,t) \quad\text{for } r \text{ on } \partial\mathcal{R} \text{ and } \Omega\cdot n(r) < 0.$$

Finite difference techniques and preconditioned iterative linear system solvers are often used for the solution of (9.22–9.23). To simplify the discussion here, however, we will consider a one-dimensional version of these equations. The difference methods used and the theoretical results established all have analogues in higher dimensions. Let $\mathcal{R}$ be the region $a < x < b$. The one-dimensional mono-energetic transport equation with isotropic scattering is

$$\frac{1}{v}\frac{\partial\psi}{\partial t} + \mu\frac{\partial\psi}{\partial x} + \sigma_t\psi - \sigma_s\phi = f, \quad x \in \mathcal{R}, \quad \mu \in [-1,1],$$

(9.24)
$$\phi(x,t) \equiv \frac{1}{2}\int_{-1}^{1}\psi(x,\mu',t)\,d\mu',$$

(9.25)
$$\psi(x,\mu,0) = \psi_0(x,\mu),$$

$$\psi(b,\mu,t) = \psi_b(\mu,t), \quad -1 \leq \mu < 0,$$

(9.26)
$$\psi(a,\mu,t) = \psi_a(\mu,t), \quad 0 < \mu \leq 1.$$

A standard approach to solving (9.24–9.26) numerically is to require that the equations hold at discrete angles μ, which are chosen to be Gauss quadrature points, and to replace the integral in (9.24) by a weighted Gauss quadrature sum. This is called the method of *discrete ordinates*:

$$\frac{1}{v}\frac{\partial\psi_j}{\partial t} + \mu_j\frac{\partial\psi_j}{\partial x} + \sigma_t\psi_j - \sigma_s\phi = f_j, \quad j = 1,\ldots,n_\mu,$$

(9.27)
$$\phi \equiv \frac{1}{2}\sum_{j'=1}^{n_\mu} w_{j'}\psi_{j'}.$$

Here ψ_j is the approximation to $\psi(x, \mu_j, t)$, and the quadrature points μ_j and weights w_j are such that for any polynomial $p(\mu)$ of degree $2n_\mu - 1$ or less,

$$\int_{-1}^{1} p(\mu') \, d\mu' = \sum_{j=1}^{n_\mu} w_j p(\mu_j).$$

We assume an even number of quadrature points n_μ so that the points μ_j are nonzero and symmetric about the origin, $\mu_{n_\mu - j+1} = -\mu_j$.

Equation (9.27) can be approximated further by a method known as *diamond differencing*—replacing derivatives in x by centered differences and approximating function values at zone centers by the average of their values at the surrounding nodes. Let the domain in x be discretized by

$$a \equiv x_0 < x_1 < \cdots < x_{n_x - 1} < x_{n_x} \equiv b,$$

and define $(\Delta x)_{i+1/2} \equiv x_{i+1} - x_i$ and $x_{i+1/2} \equiv (x_{i+1} + x_i)/2$. Equation (9.27) is replaced by

$$\frac{1}{v} \frac{\partial}{\partial t} \left(\frac{\psi_{i+1,j} + \psi_{i,j}}{2} \right) + \mu_j \frac{\psi_{i+1,j} - \psi_{i,j}}{(\Delta x)_{i+1/2}} + \sigma_t(x_{i+1/2}) \frac{\psi_{i+1,j} + \psi_{i,j}}{2}$$

$$-\sigma_s(x_{i+1/2}) \phi_{i+1/2} = f_{i+1/2,j}, \quad j = 1, \ldots, n_\mu, \quad i = 0, \ldots, n_x - 1,$$

$$(9.28) \qquad \qquad \phi_{i+1/2} \equiv \sum_{j'=1}^{n_\mu} w_{j'} \frac{\psi_{i+1,j'} + \psi_{i,j'}}{2}.$$

The combination of discrete ordinates and diamond differencing is by no means the only (or necessarily the best) technique for solving the transport equation. For a discussion of a variety of different approaches, see, for example, [93]. Still, this method is widely used, so we consider methods for solving the linear systems arising from this finite difference scheme.

Consider the time-independent version of (9.28):

$$\mu_j \frac{\psi_{i+1,j} - \psi_{i,j}}{(\Delta x)_{i+1/2}} + \sigma_t(x_{i+1/2}) \frac{\psi_{i+1,j} + \psi_{i,j}}{2} - \sigma_s(x_{i+1/2}) \phi_{i+1/2}$$

$$(9.29) \qquad = f_{i+1/2,j}, \quad \phi_{i+1/2} \equiv \sum_{j'=1}^{n_\mu} w_{j'} \frac{\psi_{i+1,j'} + \psi_{i,j'}}{2}$$

with boundary conditions

$$(9.30) \qquad \psi_{n_x,j} = \psi_b(\mu_j), \quad j \leq n_\mu/2, \qquad \psi_{0,j} = \psi_a(\mu_j), \quad j > n_\mu/2.$$

Equations (9.29–9.30) can be written in matrix form as follows. Define

$$d_{i+1/2,j} \equiv \frac{|\mu_j|}{(\Delta x)_{i+1/2}} + \frac{\sigma_t(x_{i+1/2})}{2}, \qquad e_{i+1/2,j} \equiv \frac{-|\mu_j|}{(\Delta x)_{i+1/2}} + \frac{\sigma_t(x_{i+1/2})}{2}.$$

Define $n_x + 1$-by-$n_x + 1$ triangular matrices H_j by

$$H_j \equiv \begin{pmatrix} d_{1/2,j} & e_{1/2,j} & & \\ & \ddots & & \ddots \\ & & d_{n_x-1/2,j} & e_{n_x-1/2,j} \\ & & & 1 \end{pmatrix}, \quad j \le n_\mu/2,$$

$$H_j \equiv \begin{pmatrix} 1 & & & \\ e_{1/2,j} & d_{1/2,j} & & \\ & \ddots & & \ddots \\ & & e_{n_x-1/2,j} & d_{n_x-1/2,j} \end{pmatrix}, \quad j > n_\mu/2,$$

and $n_x + 1$-by-n_x diagonal matrices $\Sigma_{s,j}$ by

$$\Sigma_{s,j} \equiv \begin{pmatrix} \sigma_{s,1/2} & & \\ & \ddots & \\ & & \sigma_{s,n_x-1/2} \\ & & 0 \end{pmatrix}, \quad j \le n_\mu/2,$$

$$\Sigma_{s,j} \equiv \begin{pmatrix} 0 & & \\ \sigma_{s,1/2} & & \\ & \ddots & \\ & & \sigma_{s,n_x-1/2} \end{pmatrix}, \quad j > n_\mu/2,$$

where $\sigma_{s,i+1/2} \equiv \sigma_s(x_{i+1/2})$. Finally, define the n_x-by-$n_x + 1$ matrix S, which averages nodal values to obtain zone-centered values, by

$$S \equiv \frac{1}{2} \begin{pmatrix} 1 & 1 & & \\ & \ddots & \ddots & \\ & & 1 & 1 \end{pmatrix}.$$

Equations (9.29–9.30) can be written in the form

$$(9.31) \quad \begin{pmatrix} H_1 & & & | & -\Sigma_{s,1} \\ & \ddots & & | & \vdots \\ & & H_{n_\mu} & | & -\Sigma_{s,n_\mu} \\ - & - & - & - & - \\ -\omega_1 S & \cdots & -\omega_{n_\mu} S & | & I \end{pmatrix} \begin{pmatrix} \psi_1 \\ \vdots \\ \psi_{n_\mu} \\ - \\ \phi \end{pmatrix} = \begin{pmatrix} f_1 \\ \vdots \\ f_{n_\mu} \\ - \\ 0 \end{pmatrix},$$

where we have taken $\omega_j \equiv w_j/2$ so that $\sum_{j=1}^{n_\mu} \omega_j = 1$, and

$$\psi_j \equiv \begin{pmatrix} \psi_{0,j} \\ \vdots \\ \psi_{n_x,j} \end{pmatrix}, \quad j = 1, \ldots, n_\mu, \quad \phi \equiv \begin{pmatrix} \phi_{1/2} \\ \vdots \\ \phi_{n_x-1/2} \end{pmatrix},$$

$$f_j \equiv \begin{pmatrix} f(x_{1/2}, \mu_j) \\ \vdots \\ f(x_{n_x-1/2}, \mu_j) \\ \psi_b(\mu_j) \end{pmatrix}, \quad j \leq n_\mu/2, \quad f_j \equiv \begin{pmatrix} \psi_a(\mu_j) \\ f(x_{1/2}, \mu_j) \\ \vdots \\ f(x_{n_x-1/2}, \mu_j) \end{pmatrix}, \quad j > n_\mu/2.$$

Usually equation (9.31) is not dealt with directly because in higher dimensions the angular flux vector ψ is quite large. The desired quantity is usually the scalar flux ϕ (from which the angular flux can be computed if needed), which is a function only of position. Therefore, the angular flux variables $\psi_1, \ldots, \psi_{n_\mu}$ are eliminated from (9.31) using Gaussian elimination, and the resulting *Schur complement* system is solved for the scalar flux ϕ:

$$(9.32) \qquad \left(I - \sum_{j=1}^{n_\mu} w_j S H_j^{-1} \Sigma_{s,j} \right) \phi = \sum_{j=1}^{n_\mu} w_j S H_j^{-1} f_j.$$

To solve this equation, one does not actually form the Schur complement matrix $A_0 \equiv I - \sum_{j=1}^{n_\mu} w_j S H_j^{-1} \Sigma_{s,j}$, which is a dense n_x-by-n_x matrix. To apply this matrix to a given vector v, one steps through each value of j, multiplying v by $\Sigma_{s,j}$, solving a triangular system with coefficient matrix H_j, multiplying the result by S, and subtracting the weighted outcome from the final vector, which has been initialized to v. In this way, only three vectors of length n_x need be stored simultaneously.

One popular method for solving equation (9.32) is to use the simple iteration defined in section 2.1 without a preconditioner; that is, the preconditioner is $M = I$. In the neutron transport literature, this is known as *source iteration*. Given an initial guess $\phi^{(0)}$, for $k = 0, 1, \ldots$, set

$$\phi^{(k+1)} = \phi^{(k)} + \sum_{j=1}^{n_\mu} w_j S H_j^{-1} f_j - A_0 \phi^{(k)}$$

$$(9.33) \qquad\qquad = \sum_{j=1}^{n_\mu} w_j S H_j^{-1} (f_j + \Sigma_{s,j} \phi^{(k)}).$$

Note that this unpreconditioned iteration for the Schur complement system (9.32) is equivalent to a preconditioned iteration for the original linear system (9.31), where the preconditioner is the *block lower triangle* of the matrix. That is, suppose $(\psi_1^{(0)}, \ldots, \psi_{n_\mu}^{(0)})^T$ is an arbitrary initial guess for the angular flux and $\phi^{(0)} = \sum_{j=1}^{n_\mu} w_j S \psi_j^{(0)}$. For $k = 0, 1, \ldots$, choose the $(k+1)$st iterate to satisfy

$$\begin{pmatrix} H_1 & & & \\ & \ddots & & \\ & & H_{n_\mu} & \\ -w_1 S & \cdots & -w_{n_\mu} S & I \end{pmatrix} \begin{pmatrix} \psi_1^{(k+1)} \\ \vdots \\ \psi_{n_\mu}^{(k+1)} \\ \phi^{(k+1)} \end{pmatrix} =$$

$$(9.34) \quad \begin{pmatrix} 0 & \cdots & 0 & \Sigma_{s,1} \\ \vdots & & \vdots & \vdots \\ 0 & \cdots & 0 & \Sigma_{s,n_\mu} \\ 0 & \cdots & 0 & 0 \end{pmatrix} \begin{pmatrix} \psi_1^{(k)} \\ \vdots \\ \psi_{n_\mu}^{(k)} \\ \phi^{(k)} \end{pmatrix} + \begin{pmatrix} f_1 \\ \vdots \\ f_{n_\mu} \\ 0 \end{pmatrix}.$$

(Equivalently, if A is the coefficient matrix in (9.31), M is the block lower triangle of A, b is the right-hand side vector in (9.31), and $u^{(k+1)}$ is the vector $(\psi_1^{(k+1)}, \ldots, \psi_{n_\mu}^{(k+1)}, \phi^{(k+1)})^T$, then $Mu^{(k+1)} = (M - A)u^{(k)} + b$ or $u^{(k+1)} = u^{(k)} + M^{-1}(b - Au^{(k)})$.) Then the scalar flux approximation at each step k satisfies

$$\phi^{(k+1)} = \sum_{j=1}^{n_\mu} w_j S \psi_j^{(k+1)} = \sum_{j=1}^{n_\mu} w_j S H_j^{-1}(f_j + \Sigma_{s,j}\phi^{(k)}),$$

which is identical to (9.33).

The coefficient matrix in (9.31) and the one in (9.32) are nonsymmetric. In general, they are not diagonally dominant, but in the special case where $e_{i+1/2,j} \leq 0$ for all i, j, the matrix in (9.31) is weakly diagonally dominant and has positive diagonal elements (since the total cross section $\sigma_t(x)$ is nonnegative and μ_j is nonzero) and nonpositive off-diagonal elements (since $\sigma_s(x) \geq 0$). We will see later that this implies certain nice properties for the block Gauss–Seidel method (9.34), such as convergence and positivity of the solution. The condition $e_{i+1/2,j} \leq 0$ is equivalent to

$$(9.35) \quad (\Delta x)_{i+1/2} \leq \frac{2|\mu_j|}{\sigma_t(x_{i+1/2})} \quad \forall i, j,$$

which means physically that the mesh width is no more than two mean free paths of the particles being simulated. It is often desirable, however, to use a coarser mesh.

Even in the more general case when (9.35) is not satisfied, it turns out that the iteration (9.33) converges. Before proving this, however, let us return to the differential equation (9.24) and use a *Fourier analysis* argument to derive an estimate of the rate of convergence that might be expected from the linear system solver. Assume that σ_s and σ_t are constant and that the problem is defined on an infinite domain. If the iterative method (9.33) is applied directly to the steady-state version of the differential equation (9.24), then we can write

$$(9.36) \quad \mu\frac{\partial \psi^{(k+1)}}{\partial x} + \sigma_t\psi^{(k+1)} = \sigma_s\phi^{(k)} + f,$$

$$(9.37) \quad \phi^{(k+1)} = \frac{1}{2}\int_{-1}^{1} \psi^{(k+1)}\, d\mu, \quad k = 0, 1, \ldots.$$

Define $\Psi^{(k+1)} \equiv \psi - \psi^{(k+1)}$ and $\Phi^{(k+1)} \equiv \phi - \phi^{(k+1)}$, where ψ, ϕ are the true solution to the steady-state version of (9.24). Then equations (9.36–9.37) give

$$(9.38) \quad \mu\frac{\partial \Psi^{(k+1)}}{\partial x} + \sigma_t\Psi^{(k+1)} = \sigma_s\Phi^{(k)},$$

$$(9.39) \qquad \Phi^{(k+1)} = \frac{1}{2} \int_{-1}^{1} \Psi^{(k+1)} \, d\mu.$$

Suppose $\Phi^{(k)}(x) = \exp(\imath\lambda x)$ and $\Psi^{(k+1)}(x, \mu) = g(\mu)\exp(\imath\lambda x)$. Introducing these expressions into equations (9.38–9.39), we find that

$$g(\mu) = \frac{\sigma_s}{\sigma_t + \imath\lambda\mu},$$

$$
\begin{aligned}
\Phi^{(k+1)}(x) &= \left(\frac{1}{2} \int_{-1}^{1} g(\mu)\, d\mu \right) \exp(\imath\lambda x) \\
&= \frac{\sigma_s}{\sigma_t} \left(\frac{1}{2} \int_{-1}^{1} \frac{1}{1 + \imath(\lambda/\sigma_t)\mu}\, d\mu \right) \exp(\imath\lambda x) \\
&= \frac{\sigma_s}{\sigma_t} \left(\frac{1}{2} \int_{-1}^{1} \frac{1}{1 + (\lambda/\sigma_t)^2\mu^2}\, d\mu \right) \Phi^{(k)}.
\end{aligned}
$$

Thus the functions $\exp(\imath\lambda x)$ are eigenfunctions of this iteration, with corresponding eigenvalues

$$\frac{\sigma_s}{\sigma_t} \left(\frac{1}{2} \int_{-1}^{1} \frac{1}{1 + (\lambda/\sigma_t)^2\mu^2}\, d\mu \right).$$

The largest eigenvalue, or spectral radius, corresponding to $\lambda = 0$ is σ_s/σ_t. Thus, we expect an asymptotic convergence rate of σ_s/σ_t.

When the iteration (9.33) is applied to the linear system (9.32), we can actually prove a stronger result. The following theorem shows that in a certain norm, the factor $\sup_x \sigma_s(x)/\sigma_t(x)$ gives a bound on the error reduction achieved at each step. Unlike the above analysis, this theorem does not require that σ_s and σ_t be constant or that the problem be defined on an infinite domain.

THEOREM 9.2.1 (Ashby et al. [3]). *Assume* $\sigma_t(x) \geq \sigma_s(x) \geq 0$ *for all* $x \in \mathcal{R} \equiv (a, b)$, *and assume also that* $\sigma_t(x) \geq c > 0$ *on* $\mathcal{R}$. *Then for each* j,

$$(9.40) \qquad \|\Theta^{1/2} S H_j^{-1} \Sigma_{s,j} \Theta^{-1/2}\| < \sup_{x \in \mathcal{R}} \sigma_s(x)/\sigma_t(x) \leq 1,$$

where $\Theta = \operatorname{diag}(\sigma_t(x_{1/2})(\Delta x)_{1/2}, \ldots, \sigma_t(x_{n_x-1/2})(\Delta x)_{n_x-1/2})$.

Proof. First note that the n_x-by-n_x matrix $S H_j^{-1} \Sigma_{s,j}$ is the same matrix obtained by taking the product of the upper left n_x-by-n_x blocks of S, H_j^{-1}, and $\Sigma_{s,j}$ for $j \leq n_\mu/2$ or the lower right n_x-by-n_x blocks of $S\, H_j^{-1}$, and $\Sigma_{s,j}$ for $j > n_\mu/2$. Accordingly, let $\hat{S}_j$, $\hat{H}_j^{-1}$, and $\hat{\Sigma}_{s,j}$ denote these n_x-by-n_x blocks. We will establish the bound (9.40) for $\|\Theta^{1/2} \hat{S}_j \hat{H}_j^{-1} \hat{\Sigma}_{s,j} \Theta^{-1/2}\|$.

Note that $\hat{H}_j$ can be written in the form

$$\hat{H}_j = \hat{D}_j \hat{G}_j + \hat{\Sigma}_t \hat{S}_j,$$

where

$$\hat{D}_j = \begin{pmatrix} \frac{|\mu_j|}{(\Delta x)_{1/2}} & & \\ & \ddots & \\ & & \frac{|\mu_j|}{(\Delta x)_{n_x-1/2}} \end{pmatrix}, \quad \hat{\Sigma}_t = \begin{pmatrix} \sigma_{t,1/2} & & \\ & \ddots & \\ & & \sigma_{t,n_x-1/2} \end{pmatrix},$$

$$\hat{G}_j = \begin{pmatrix} 1 & -1 & & \\ & \ddots & \ddots & \\ & & 1 & -1 \\ & & & 1 \end{pmatrix}, \quad j \le n_\mu/2, \quad \hat{G}_j = \begin{pmatrix} 1 & & & \\ -1 & 1 & & \\ & \ddots & \ddots & \\ & & -1 & 1 \end{pmatrix}, \quad j > n_\mu/2.$$

It can be seen that for each j, $\hat{G}_j = 2(I - \hat{S}_j)$. Dropping the subscript j for convenience, we can write

$$\begin{aligned}
\hat{S}\hat{H}^{-1}\hat{\Sigma}_s &= \hat{S}\left(2\hat{D}(I - \hat{S}) + \hat{\Sigma}_t\hat{S}\right)^{-1}\hat{\Sigma}_s \\
&= \left(2\hat{D}(\hat{S}^{-1} - I) + \hat{\Sigma}_t\right)^{-1}\hat{\Sigma}_s \\
&= \left(2\hat{\Sigma}_t^{-1}\hat{D}(\hat{S}^{-1} - I) + I\right)^{-1}\hat{\Sigma}_t^{-1}\hat{\Sigma}_s.
\end{aligned}$$

Multiplying by $\Theta^{1/2}$ on the left and by $\Theta^{-1/2}$ on the right gives

$$\Theta^{1/2}\hat{S}\hat{H}^{-1}\hat{\Sigma}\Theta^{-1/2} = \left(2|\mu_j|\Theta^{-1/2}(\hat{S}^{-1} - I)\Theta^{-1/2} + I\right)^{-1}\hat{\Sigma}_t^{-1}\hat{\Sigma},$$

and it follows that

$$\|\Theta^{1/2}\hat{S}\hat{H}^{-1}\hat{\Sigma}\Theta^{-1/2}\| \le \left\|\left(2|\mu_j|\Theta^{-1/2}(\hat{S}^{-1} - I)\Theta^{-1/2} + I\right)^{-1}\right\| \cdot \gamma,$$

$$(9.41) \qquad \gamma \equiv \|\hat{\Sigma}_t^{-1}\hat{\Sigma}_s\| \le \sup_{x \in \mathcal{R}} \sigma_s(x)/\sigma_t(x).$$

The matrix norm on the right-hand side in (9.41) is equal to the inverse of the square root of the smallest eigenvalue of

$$I + 2|\mu_j|\Theta^{-1/2}(\hat{S}^{-1} + \hat{S}^{-T} - 2I)\Theta^{-1/2}$$

$$(9.42) \qquad +4|\mu_j|^2\Theta^{-1/2}(\hat{S}^{-1} - I)^T\Theta^{-1}(\hat{S}^{-1} - I)\Theta^{-1/2}.$$

The third term in this sum is positive definite, since

$$\hat{S}^{-1} - I = \begin{pmatrix} 1 & -2 & \cdots & (-1)^{n_x+1}2 \\ & \ddots & \ddots & \vdots \\ & & 1 & (-1)^{2n_x-1}2 \\ & & & 1 \end{pmatrix}$$

is nonsingular and $|\mu_j| > 0$. It will follow that the smallest eigenvalue of the matrix in (9.42) is strictly greater than 1 if the second term,

$$(9.43) \qquad 2|\mu_j|\Theta^{-1/2}(\hat{S}^{-1} + \hat{S}^{-T} - 2I)\Theta^{-1/2},$$

can be shown to be positive semidefinite. Directly computing this matrix, we find that

$$\hat{S}^{-1} + \hat{S}^{-T} - 2I = 2 \begin{pmatrix} 1 & -1 & \cdots & (-1)^{n_x+1} \\ -1 & 1 & \cdots & (-1)^{n_x+2} \\ \vdots & \vdots & & \vdots \\ (-1)^{n_x+1} & (-1)^{n_x+2} & \cdots & 1 \end{pmatrix},$$

which has $n_x - 1$ eigenvalues equal to 0 and the remaining eigenvalue equal to $2n_x$. Hence this matrix and the one in (9.43) are positive semidefinite. It follows that the matrix norm on the right-hand side in (9.41) is strictly less than 1, and from this the desired result is obtained. $\quad\square$

COROLLARY 9.2.1. *Under the assumptions of Theorem 9.2.1, the iteration (9.33) converges to the solution ϕ of (9.32), and if $e^{(k)} \equiv \phi - \phi^{(k)}$ denotes the error at step k, then*

$$(9.44) \qquad \|\Theta^{1/2} e^{(k+1)}\| < \gamma \, \|\Theta^{1/2} e^{(k)}\|,$$

where Θ is defined in the theorem and γ is defined in (9.41).

Proof. From (9.33) we have

$$\Theta^{1/2} e^{(k+1)} = \sum_{j=1}^{n_\mu} \omega_j (\Theta^{1/2} S H_j^{-1} \Sigma_j \Theta^{-1/2}) \Theta^{1/2} e^{(k)},$$

and taking norms on both sides and recalling that the weights ω_j are nonnegative and sum to 1, we find

$$\|\Theta^{1/2} e^{(k+1)}\| < \sum_{j=1}^{n_\mu} \omega_j \gamma \|\Theta^{1/2} e^{(k)}\| = \gamma \|\Theta^{1/2} e^{(k)}\|.$$

Since $\gamma \leq 1$ and since the inequality in (9.44) is strict, with the amount by which the actual reduction factor differs from γ being independent of k, it follows that the iteration (9.33) converges to the solution of (9.32). $\quad\square$

For $\gamma \ll 1$, Corollary 9.2.1 shows that the simple source iteration (9.33) converges rapidly, but for $\gamma \approx 1$, convergence may be slow. In Part I of this book, we discussed many ways to accelerate the simple iteration method, such as Orthomin(1), QMR, BiCGSTAB, or full GMRES. Figure 9.2 shows the convergence of simple iteration, Orthomin(1), and full GMRES applied to two test problems. QMR and BiCGSTAB were also tested on these problems, and each required only slightly more iterations than full GMRES, but at twice the cost in terms of matrix–vector multiplications. The vertical axis is the ∞-norm of the error in the approximate solution. The exact solution to the linear system was computed directly for comparison. Here we used a uniform mesh spacing $\Delta x = .25$ ($n_x = 120$) and eight angles, but the convergence rate was not very sensitive to these mesh parameters.

The first problem, taken from [92], is a model shielding problem, with cross sections corresponding to water and iron in different regions, as illustrated below.

water	water	iron	water
$0 \le x \le 12$	$12 \le x \le 15$	$15 \le x \le 21$	$21 \le x \le 30$
$\sigma_t = 3.3333$	$\sigma_t = 3.3333$	$\sigma_t = 1.3333$	$\sigma_t = 3.3333$
$\sigma_s = 3.3136$	$\sigma_s = 3.3136$	$\sigma_s = 1.1077$	$\sigma_s = 3.3136$
$f = 1$	$f = 0$	$f = 0$	$f = 0$

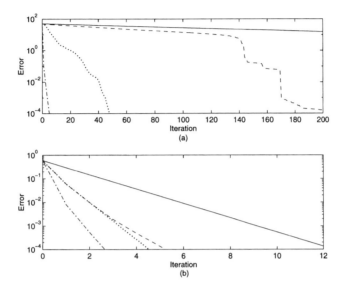

FIG. 9.2.   *Error curves for* (a) $\gamma = .994$ *and* (b) $\gamma = .497$. *Simple iteration (solid), Orthomin(1) (dashed), full GMRES (dotted), and DSA-preconditioned simple iteration (dash–dot).*

The slab thicknesses are in cm and the cross sections are in cm^{-1}. There is a vacuum boundary condition at the right end ($\psi_{n_x,j} = 0$, $j \le n_\mu/2$) and a reflecting boundary condition at the left ($\psi_{0,n_\mu-j+1} = \psi_{0,j}$, $j \le n_\mu/2$). A uniform source $f = 1$ is placed in the first (leftmost) region. In the second test problem, we simply replaced σ_s in each region by half its value: $\sigma_s = 1.6568$ in the first, second, and fourth regions; $\sigma_s = .55385$ in the third.

Also shown in Figure 9.2 is the convergence of the simple iteration method with a preconditioner designed specifically for the transport equation known as *diffusion synthetic acceleration (DSA)*. In the first problem, where $\gamma = .994$, it is clear that the unpreconditioned simple iteration (9.33) is unacceptably slow to converge. The convergence rate is improved significantly by Orthomin(1), with little extra work and storage per iteration, and it is improved even more by full GMRES but at the cost of extra work and storage. The most effective method for solving this problem, however, is the DSA-preconditioned simple iteration.

For the second problem, the reduction in the number of iterations is less dramatic. (Note the different horizontal scales in the two graphs.) Unpre-conditioned simple iteration converges fairly rapidly, Orthomin(1) reduces the number of iterations by about a factor of 2, and further accelerations such as full GMRES and DSA can bring about only a modest reduction in the number of iteration steps. If the cost of an iteration is significantly greater, these methods will not be cost effective.

For the time-dependent problem (9.28), the time derivative term essentially adds to the total cross section σ_t. That is, suppose (9.28) is solved using centered differences in time. The equations for $\psi_{i,j}^{\ell+1}$ at time $t_{\ell+1} = t_\ell + \Delta t$ become

$$\frac{1}{v} \frac{\psi_{i+1,j}^{\ell+1} + \psi_{i,j}^{\ell+1}}{(\Delta t)} + \mu_j \frac{\psi_{i+1,j}^{\ell+1} - \psi_{i,j}^{\ell+1}}{(\Delta x)_{i+1/2}} + \sigma_{t,i+1/2} \frac{\psi_{i+1,j}^{\ell+1} + \psi_{i,j}^{\ell+1}}{2}$$

$$-\sigma_{s,i+1/2}\phi_{i+1/2}^{\ell+1} = 2f_{i+1/2,j} + \frac{1}{v} \frac{\psi_{i+1,j}^{\ell} + \psi_{i,j}^{\ell}}{(\Delta t)}$$

$$- \left[\mu_j \frac{\psi_{i+1,j}^{\ell} - \psi_{i,j}^{\ell}}{(\Delta x)_{i+1/2}} + \sigma_{t,i+1/2} \frac{\psi_{i+1,j}^{\ell} + \psi_{i,j}^{\ell}}{2} - \sigma_{s,i+1/2}\phi_{i+1/2}^{\ell} \right].$$

The matrix equation for the flux $\psi^{\ell+1}$ in terms of f and ψ^ℓ is like that in (9.31), except that the entries $d_{i+1/2,j}$ and $e_{i+1/2,j}$ of H_j are each increased by $1/(v\Delta t)$. One would obtain the same coefficient matrix for the steady-state problem if σ_t were replaced by $\sigma_t + 2/(v\Delta t)$. Thus, for time-dependent problems the convergence rate of iteration (9.33) at time step $\ell+1$ is governed by the quantity

$$\sup_{x \in \mathcal{R}} \frac{\sigma_s(x, t_{\ell+1})}{\sigma_t(x, t_{\ell+1}) + 2/(v\Delta t)}.$$

In many cases, this quantity is bounded well away from 1, even if σ_s/σ_t is not.

For steady-state problems with $\gamma \approx 1$, it is clear from Figure 9.2a that the DSA preconditioner is extremely effective in terms of reducing the number of iterations. At each iteration a linear system corresponding to the steady-state diffusion equation must be solved. Since this is a book on iterative methods and not specifically on the transport equation, we will not give a complete account of diffusion synthetic acceleration. For a discussion and analysis, see [91, 3]. The basic idea, however, is that when $\sigma_s/\sigma_t \approx 1$, the scalar flux ϕ approximately satisfies a diffusion equation and therefore the diffusion operator is an effective preconditioner for the linear system. In one dimension, the diffusion operator is represented by a tridiagonal matrix which is easy to solve, but in higher dimensions, the diffusion equation itself may require an iterative solution technique. The advantage of solving the diffusion equation is that it is independent of angle. An iteration for the diffusion equation requires about $1/n_\mu$ times as much work as an iteration for equation (9.32), so a number of inner iterations on the preconditioner may be acceptable in order to reduce the number of outer iterations. Of course, the diffusion operator could be used as a preconditioner for other iterative methods as well; that is, DSA could be further accelerated by replacing the simple iteration strategy with, say, GMRES.

A number of different formulations and solution techniques for the transport equation have been developed. In [96, 97], for example, multigrid methods are applied to the transport equation. The development of accurate and efficient methods for solving the transport equation remains an area of active research.

Comments and Additional References.

Iterative solution of the transport equation requires at least two levels of nested iterations—an outer iteration over energy groups and an inner iteration for each group. If the DSA preconditioner is used, then a third level of iteration may be required to solve the diffusion equation. Since the ultimate goal is to solve the outermost linear system, one might consider accepting less accurate solutions to the inner linear systems, especially at early stages of the outer iteration, if this would lead to less total work in solving the outermost system.

The appropriate level of accuracy depends, of course, on the iterative methods used. As might be guessed from the analysis of Chapter 4, the CG method is especially sensitive to errors (rounding errors or otherwise), so an outer CG iteration may require more accuracy from an inner iteration. This might be a motivation for using a different outer iteration, such as the Chebyshev method [61, 94]. (Of course, the transport equation, in the form stated in this chapter, is nonsymmetric, so the CG method could not be used anyway, unless it was applied to the normal equations.)

For discussions of accuracy requirements in inner and outer iterations, see [59, 56].

Exercises.

9.1. Use the Taylor series to show that the approximation

$$\left(\frac{\partial}{\partial x} a \frac{\partial u}{\partial x}\right)(x_i, y_j) \approx \frac{a_{i+1/2,j}(u_{i+1,j} - u_{i,j}) - a_{i-1/2,j}(u_{i,j} - u_{i-1,j})}{h_x^2}$$

is second-order accurate, provided that $a \partial u / \partial x \in \mathbf{C}^3$ and $a \partial^3 u / \partial x^3 \in \mathbf{C}^1$; that is, show that the absolute value of the difference between the right- and left-hand sides is bounded by

$$\frac{h_x^2}{24} \left[\max_{x \in [x_{i-1}, x_{i+1}]} \left| \frac{\partial^3}{\partial x^3} \left(a \frac{\partial u}{\partial x} \right) \right| + \max_{x \in [x_{i-1}, x_{i+1}]} \left| \frac{\partial}{\partial x} \left(a \frac{\partial^3 u}{\partial x^3} \right) \right| \right].$$

9.2. Let $u(x, y)$ be the solution to Poisson's equation $\nabla^2 u = f$ on the unit square with homogeneous Dirichlet boundary conditions: $u(x, 0) = u(x, 1) = u(0, y) = u(1, y) = 0$, and let $\mathbf{u}$ be the vector of values $u(x_i, y_j)$ on a uniform grid of spacing h in each direction. Let $\hat{u}$ be the solution to the linear system $\nabla_h^2 \hat{u} = \mathbf{f}$, where ∇_h^2 represents the matrix defined in (9.10–9.11), with $h_x = h_y = h$, and $\mathbf{f}$ is the vector of right-hand side values $f(x_i, y_j)$. Use the previous exercise and Corollary 9.1.2 to show that

$$(9.45) \qquad \sqrt{\frac{1}{n} \sum_{i=1}^{n} (u_i - \hat{u}_i)^2} \leq Ch^2$$

for some constant C independent of h. (Note that the ordinary Euclidean norm of the difference between $\mathbf{u}$ and $\hat{u}$ is not $O(h^2)$ but only $O(h)$. The norm in (9.45) is more like the $\mathcal{L}_2$ norm for *functions*:

$$\|g\|_{\mathcal{L}_2} \equiv \left(\int_0^1 \int_0^1 g^2(x, y) \, dx \, dy \right)^{1/2}.$$

This is a reasonable way to measure the error in a vector that approximates a function at n points, since if the difference is equal to ϵ at each point, then the error norm in (9.45) is ϵ, not $\sqrt{n}\epsilon$.)

9.3. Show that the eigenvectors in (9.18) are orthonormal.

9.4. Use Theorem 9.2.1 to show that if Orthomin(1) is applied to the scaled transport equation

$$\Theta^{1/2} \left(I - \sum_{j=1}^{n_\mu} w_j SH_j^{-1} \Sigma_{s,j} \right) \Theta^{-1/2} \hat{\phi} = \Theta^{1/2} \sum_{j=1}^{n_\mu} w_j SH_j^{-1} f_j,$$

where $\phi = \Theta^{-1/2}\hat{\phi}$, then it will converge to the solution for any initial vector, and, at each step, the 2-norm of the residual (in the scaled equation) will be reduced by at least the factor γ in (9.41).

9.5. A physicist has a code that solves the transport equation using source iteration (9.33). She decides to improve the approximation by replacing $\phi^{(k+1)}$ at each step with the linear combination $\alpha_{k+1}\phi^{(k+1)} + (1 - \alpha_{k+1})\phi^{(k)}$, where α_{k+1} is chosen to make the 2-norm of the residual as small as possible. Which of the methods described in this book is she using?

Chapter 10

Comparison of Preconditioners

We first briefly consider the classical iterative methods—Jacobi, Gauss–Seidel, and SOR. Then more general theory is developed for comparing preconditioners used with simple iteration or with the conjugate gradient or MINRES methods for symmetric positive definite problems. Most of the theorems in this chapter (and throughout the remainder of this book) apply only to *real* matrices, but this restriction will be apparent from the hypotheses of the theorem. The algorithms can be used for complex matrices as well.

10.1. Jacobi, Gauss–Seidel, SOR.

An equivalent way to describe Algorithm 1 of section 2.1 is as follows. Write A in the form $A = M - N$ so that the linear system $Ax = b$ becomes

$$(10.1) \qquad\qquad Mx = Nx + b.$$

Given an approximation x_{k-1}, obtain a new approximation x_k by substituting x_{k-1} into the right-hand side of (10.1) so that

$$(10.2) \qquad\qquad Mx_k = Nx_{k-1} + b.$$

To see that (10.2) is equivalent to Algorithm 1, multiply by M^{-1} in (10.2) and substitute $M^{-1}N = I - M^{-1}A$ to obtain

$$x_k = (I - M^{-1}A)x_{k-1} + M^{-1}b = x_{k-1} + M^{-1}r_{k-1} = x_{k-1} + z_{k-1}.$$

The simple iteration algorithm was traditionally described by (10.2), and the decomposition $A = M - N$ was referred to as a *matrix splitting*. The terms "matrix splitting" and "preconditioner," when referring to the matrix M, are synonymous.

If M is taken to be the *diagonal* of A, then the simple iteration procedure with this matrix splitting is called *Jacobi's method*. We assume here that the diagonal entries of A are nonzero, so M^{-1} is defined. It is sometimes useful to write the matrix equation (10.2) in element form to see exactly how the update to the approximate solution vector is accomplished. Using parentheses to denote components of vectors, Jacobi's method can be written in the form

$$(10.3) \qquad x_k(i) = \frac{1}{a_{ii}} \left(-\sum_{j \neq i} a_{ij} x_{k-1}(j) + b(i) \right), \quad i = 1, \dots, n.$$

Note that the new vector x_k cannot overwrite x_{k-1} in Jacobi's method until all of its entries have been computed.

If M is taken to be the *lower triangle* of A, then the simple iteration procedure is called the *Gauss–Seidel method*. Equations (10.2) become

$$(10.4) \quad x_k(i) = \frac{1}{a_{ii}} \left(-\sum_{j=1}^{i-1} a_{ij} x_k(i) - \sum_{j=i+1}^{n} a_{ij} x_{k-1}(j) + b(i) \right), \quad i = 1, \dots, n.$$

For the Gauss–Seidel method, the latest approximations to the components of x are used in the update of subsequent components. It is convenient to overwrite the old components of x_{k-1} with those of x_k as soon as they are computed.

The convergence rate of the Gauss–Seidel method often can be improved by introducing a *relaxation parameter* ω. The *SOR* (successive overrelaxation) method is defined by

$$x_k(i) = \omega \frac{1}{a_{ii}} \left(-\sum_{j=1}^{i-1} a_{ij} x_k(j) - \sum_{j=i+1}^{n} a_{ij} x_{k-1}(j) + b(i) \right)$$

$$(10.5) \qquad\qquad + (1-\omega) x_{k-1}(i), \quad i = 1, \dots, n.$$

In matrix form, if $A = D - L - U$, where D is diagonal, L is strictly lower triangular, and U is strictly upper triangular, then $M = \omega^{-1} D - L$. The method should actually be called overrelaxation or underrelaxation, according to whether $\omega > 1$ or $\omega < 1$. When $\omega = 1$ the SOR method reduces to Gauss–Seidel. In the Gauss–Seidel method, each component $x_k(i)$ is chosen so that the ith equation is satisfied by the current partially updated approximate solution vector. For the SOR method, the ith component of the current residual vector is $(1-\omega) a_{ii}(\tilde{x}_k(i) - x_{k-1}(i))$, where $\tilde{x}_k(i)$ is the value that would make the ith component of the residual zero.

Block versions of the Jacobi, Gauss–Seidel, and SOR iterations are easily defined. (Here we mean block preconditioners, not blocks of iteration vectors as in section 7.4.) If M is taken to be the block diagonal of A—that is, if A is of the form

$$A = \begin{pmatrix} A_{1,1} & A_{1,2} & \cdots & A_{1,m} \\ A_{2,1} & A_{2,2} & \cdots & A_{2,m} \\ \vdots & \vdots & & \vdots \\ A_{m,1} & A_{m,2} & \cdots & A_{m,m} \end{pmatrix},$$

where each diagonal block $A_{i,i}$ is square and nonsingular, and

$$M = \begin{pmatrix} A_{1,1} & & \\ & \ddots & \\ & & A_{m,m} \end{pmatrix}$$

—then the simple iteration procedure with this matrix splitting is called the block Jacobi method. Similarly, for M equal to the block lower triangle of A, we obtain the block Gauss–Seidel method; for M of the form $\omega^{-1}D - L$, where D is the block diagonal and L is the strictly block lower triangular part of A, we obtain the block SOR method.

When A is real symmetric or complex Hermitian, then symmetric or Hermitian versions of the Gauss–Seidel and SOR preconditioners can be defined. If one defines $M_1 = \omega^{-1}D - L$, as in the SOR method, and $M_2 = \omega^{-1}D - U$ and sets

$$M_1 x_{k-1/2} = N_1 x_{k-1} + b, \quad N_1 \equiv M_1 - A,$$
$$M_2 x_k = N_2 x_{k-1/2} + b, \quad N_2 \equiv M_2 - A,$$

then the resulting iteration is known as the symmetric SOR or SSOR method. It is left as an exercise to show that the preconditioner M in this case is

$$(10.6) \qquad M = \frac{\omega}{2 - \omega}(\omega^{-1}D - L)D^{-1}(\omega^{-1}D - U);$$

that is, if we eliminate $x_{k-1/2}$, then x_k satisfies $M x_k = N x_{k-1} + b$, where $A = M - N$. The SSOR preconditioner is sometimes used with the CG algorithm for Hermitian positive definite problems.

10.1.1. Analysis of SOR.
A beautiful theory describing the convergence rate of the SOR iteration and the optimal value for ω was developed by Young [144]. We include here only the basics of that theory, for two reasons. First, it is described in many other places. In addition to [144], see, for instance, [77, 83].

Second, the upshot of the theory is an expression for the optimal value of ω and the spectral radius of the SOR iteration matrix in terms of the spectral radius of the Jacobi iteration matrix. In most practical applications, the spectral radius of the Jacobi matrix is not known, so computer programs have been developed to try to dynamically estimate the optimal value of ω. The theory is most often applied to the model problem for Poisson's equation on a square, described in section 9.1.1, because here the eigenvalues are known. For this problem it can be shown that with the optimal value of ω, the spectral radius of the SOR iteration matrix is $1 - O(h)$ instead of $1 - O(h^2)$ as it is for $\omega = 1$. This is a tremendous improvement; it means that the number of iterations required to achieve a fixed level of accuracy is reduced from $O(h^{-2})$ to $O(h^{-1})$. (Recall that the spectral radius is the same as the 2-norm for a

Hermitian matrix. Hence it determines not only the asymptotic convergence rate but the amount by which the error is reduced at each step.)

One obtains the same level of improvement, however, with the unpreconditioned CG algorithm, and here there are no parameters to estimate. The development of the CG algorithm for Hermitian positive definite problems has made SOR theory less relevant. Therefore, we will concentrate most of our effort on finding preconditioners that lead to still further improvement on this $O(h^{-1})$ estimate.

Let A be written in the form $A = D - L - U$, where D is diagonal, L is strictly lower triangular, and U is strictly upper triangular. The asymptotic convergence rates of the Jacobi and SOR methods depend on the spectral radii of $G_J \equiv I - D^{-1}A = D^{-1}(L + U)$ and $G_\omega \equiv I - (\omega^{-1}D - L)^{-1}A = (D - \omega L)^{-1}[(1 - \omega)D + \omega U]$, respectively. Note that if we prescale A by its diagonal so that $\tilde{A} \equiv D^{-1}A = I - D^{-1}L - D^{-1}U$, then the Jacobi and SOR iteration matrices do not change. For convenience, let us assume that A has been prescaled by its diagonal and let L and U now denote the strictly lower and strictly upper triangular parts of the scaled matrix, $A = I - L - U$. Then the Jacobi and SOR iteration matrices are

$$G_J \equiv I - A = L + U \quad \text{and}$$

(10.7) $$G_\omega \equiv I - (\omega^{-1}I - L)^{-1}A = (I - \omega L)^{-1}[(1 - \omega)I + \omega U].$$

We first note that for the SOR method, we need only consider values of ω in the open interval $(0, 2)$.

THEOREM 10.1.1. *For any $\omega \in \mathbf{C}$, we have*

(10.8) $$\rho(G_\omega) \geq |1 - \omega|.$$

Proof. From (10.7) it follows that

$$\begin{aligned}
\det(G_\omega) &= \det[(I - \omega L)^{-1}[(1 - \omega)I + \omega U]] \\
&= \det[(I - \omega L)^{-1}] \cdot \det[(1 - \omega)I + \omega U].
\end{aligned}$$

Since the matrices here are triangular, their determinants are equal to the product of their diagonal entries, so we have $\det(G_\omega) = (1 - \omega)^n$. The determinant of G_ω is also equal to the product of its eigenvalues, and it follows that at least one of the n eigenvalues must have absolute value greater than or equal to $|1 - \omega|$.   □

Theorem 10.1.1 holds for any matrix A (with nonzero diagonal entries). By making additional assumptions about the matrix A, one can prove more about the relation between the convergence rates of the Jacobi, Gauss–Seidel, and SOR iterations. In the following theorems, we make what seems to be a rather unusual assumption (10.9). We subsequently note that this assumption can sometimes be verified just by considering the sparsity pattern of A.

THEOREM 10.1.2. *Suppose that the matrix $A = I - L - U$ has the following property: for any $c \in \mathbf{R}$,*

(10.9) $$det(cI - L - U) = det(cI - \gamma L - \gamma^{-1}U)$$

for all $\gamma \in \mathbf{R} \backslash \{0\}$. Then the following properties hold:

(i) *If μ is an eigenvalue of G_J, then $-\mu$ is an eigenvalue of G_J with the same multiplicity.*

(ii) *If $\lambda = 0$ is an eigenvalue of G_ω, then $\omega = 1$.*

(iii) *If $\lambda \neq 0$ is an eigenvalue of G_ω for some $\omega \in (0, 2)$, then*

(10.10) $$\mu = \frac{\lambda + \omega - 1}{\omega \lambda^{1/2}}$$

is an eigenvalue of G_J.

(iv) *If μ is an eigenvalue of G_J and λ satisfies (10.10) for some $\omega \in (0, 2)$, then λ is an eigenvalue of G_ω.*

Proof. From property (10.9) with $\gamma = -1$, we have, for any number μ,

$$det(G_J - \mu I) = det(L + U - \mu I) = det(-L - U - \mu I)$$
$$= (-1)^n det(L + U + \mu I) = (-1)^n det(G_J + \mu I).$$

Since the eigenvalues of G_J are the numbers μ for which $det(G_J - \mu I) = 0$ and their multiplicities are also determined by this characteristic polynomial, result (i) follows.

Since the matrix $I - \omega L$ is lower triangular with ones on the diagonal, its determinant is 1; for any number λ we have

$$
\begin{aligned}
det(G_\omega - \lambda I) &= det[(I - \omega L)^{-1}[(1 - \omega)I + \omega U] - \lambda I] \\
&= det[(1 - \omega)I + \omega U - \lambda(I - \omega L)].
\end{aligned}
$$
(10.11)

If $\lambda = 0$ is an eigenvalue of G_ω, then (10.11) implies that $det[(1-\omega)I+\omega U)] = 0$. Since this matrix is upper triangular with $(1 - \omega)$'s along the diagonal, we deduce that $\omega = 1$ and thus prove (ii).

For $\lambda \neq 0$, equation (10.11) implies that

$$det(G_\omega - \lambda I) = \omega^n \lambda^{n/2} det\left[\frac{1 - \omega - \lambda}{\omega \lambda^{1/2}} I + \lambda^{-1/2}U + \lambda^{1/2}L\right].$$

Using property (10.9) with $\gamma = \lambda^{-1/2}$, we have

(10.12) $$det(G_\omega - \lambda I) = \omega^n \lambda^{n/2} det\left[G_J - \frac{\lambda + \omega - 1}{\omega \lambda^{1/2}}I\right].$$

It follows that if $\lambda \neq 0$ is an eigenvalue of G_ω and if μ satisfies (10.10), then μ is an eigenvalue of G_J. Conversely, if μ is an eigenvalue of G_J and λ satisfies (10.10), then λ is an eigenvalue of G_ω. This proves (iii) and (iv).   □

COROLLARY 10.1.1. *When the coefficient matrix A satisfies (10.9), asymptotically the Gauss–Seidel iteration is twice as fast as the Jacobi iteration; that is, $\rho(G_1) = (\rho(G_J))^2$.*

Proof. For $\omega = 1$, (10.10) becomes

$$\mu = \lambda^{1/2}.$$

If all eigenvalues λ of G_1 are 0, then part (iv) of Theorem 10.1.2 implies that all eigenvalues of G_J are 0 as well. If there is a nonzero eigenvalue λ of G_1, then part (iii) of Theorem 10.1.2 implies that there is an eigenvalue μ of G_J such that $\mu = \lambda^{1/2}$. Hence $\rho(G_J)^2 \geq \rho(G_1)$. Part (iv) of Theorem 10.1.2 implies that there is no eigenvalue μ of G_J such that $|\mu|^2 > \rho(G_1)$; if there were such an eigenvalue μ, then $\lambda = \mu^2$ would be an eigenvalue of G_1, which is a contradiction. Hence $\rho(G_J)^2 = \rho(G_1)$. $\square$

In some cases—for example, when A is Hermitian—the Jacobi iteration matrix G_J has only *real* eigenvalues. The SOR iteration matrix is non-Hermitian and may well have complex eigenvalues, but one can prove the following theorem about the optimal value of ω for the SOR iteration and the corresponding optimal convergence rate.

THEOREM 10.1.3. *Suppose that A satisfies (10.9), that G_J has only* real *eigenvalues, and that $\beta \equiv \rho(G_J) < 1$. Then the SOR iteration converges for every $\omega \in (0, 2)$, and the spectral radius of the SOR matrix is*

$$(10.13) \quad \rho(G_\omega) = \begin{cases} \frac{1}{4}\left[\omega\beta + \sqrt{(\omega\beta)^2 - 4(\omega - 1)}\right]^2 & \text{for } 0 < \omega \leq \omega_{opt}, \\ \omega - 1 & \text{for } \omega_{opt} \leq \omega < 2, \end{cases}$$

where ω_{opt}, the optimal value of ω, is

$$(10.14) \qquad\qquad \omega_{opt} = \frac{2}{1 + \sqrt{1 - \beta^2}}.$$

For any other value of ω, we have

$$(10.15) \qquad\qquad \rho(G_{\omega_{opt}}) < \rho(G_\omega), \quad \omega \in (0, 2)\backslash\{\omega_{opt}\}.$$

Proof. Solving (10.10) for λ gives

$$(10.16) \qquad\qquad \lambda = \frac{1}{4}\left(\omega\mu \pm \sqrt{(\omega\mu)^2 - 4(\omega - 1)}\right)^2.$$

It follows from Theorem 10.1.2 that if μ is an eigenvalue of G_J, then both roots λ are eigenvalues of G_ω.

Since μ is real, the term inside the square root in (10.16) is negative if

$$\tilde{\omega} \equiv \frac{2(1 - \sqrt{1 - \mu^2})}{\mu^2} < \omega < 2,$$

and in this case

$$
\begin{aligned}
|\lambda| &= \frac{1}{4}\left[(\omega\mu)^2 + 4(\omega - 1) - (\omega\mu)^2\right] \\
&= \omega - 1, \quad \omega \in (\tilde{\omega}, 2).
\end{aligned}
$$

(10.17)

In the remaining part of the range of ω, both roots λ are positive and the larger one is

(10.18) $\qquad \dfrac{1}{4}\left[\omega|\mu| + \sqrt{(\omega|\mu|)^2 - 4(\omega - 1)}\right]^2, \quad \omega \in (0, \tilde{\omega}].$

Also, this value is greater than or equal to $\omega - 1$ for $\omega \in (0, \tilde{\omega}]$ since in this range we have

$$
\frac{1}{4}\left[\omega|\mu| + \sqrt{(\omega|\mu|)^2 - 4(\omega - 1)}\right]^2 \geq \frac{1}{4}(\omega|\mu|)^2 \geq \omega - 1.
$$

It is easy to check that for any fixed $\omega \in (0, \tilde{\omega}]$, expression (10.18) is a strictly increasing function of $|\mu|$. Likewise, $\tilde{\omega}$ is a strictly increasing function of $|\mu|$, and we have

$$
\tilde{\omega} \leq \frac{2(1 - \sqrt{1 - \beta^2})}{\beta^2} = \frac{2}{1 + \sqrt{1 - \beta^2}} \equiv \omega_{opt}.
$$

It follows that an eigenvalue λ of G_ω for which $|\lambda| = \rho(G_\omega)$ corresponds to an eigenvalue μ of G_J for which $|\mu| = \beta$ because such an eigenvalue is greater than or equal to those corresponding to smaller values of $|\mu|$ if $\omega \in (0, \omega_{opt}]$, and it is equal to the others if $\omega \in (\omega_{opt}, 2)$. We thus deduce that (10.13) holds for ω_{opt} given by (10.14). Since the expressions in (10.13) are less than 1 for all $\omega \in (0, 2)$, the SOR iteration converges. It can also be seen that for fixed $|\mu| = \beta$, expression (10.18) is a strictly decreasing function of ω for $\omega \in (0, \omega_{opt}]$, thereby reaching its minimum at $\omega = \omega_{opt}$. Inequality (10.15) is then proved. $\square$

The expression in (10.13) for $\rho(G_\omega)$ is plotted in Figure 10.1 for different values of $\beta = \rho(G_J)$. It can be seen from the figure that if the optimal value ω_{opt} is not known, then it is better to overestimate it than to underestimate it, especially for values of β near 1. Some computer codes have been designed to estimate ω_{opt} dynamically, but these will not be discussed here.

The condition (10.9) of Theorem 10.1.2 can sometimes be established just by considering the sparsity pattern of A.

DEFINITION 10.1.1. *A matrix A of order n has* Property A *if there exist two disjoint subsets S_1 and S_2 of $\mathbf{Z}^n = \{1, \ldots, n\}$ such that $S_1 \bigcup S_2 = \mathbf{Z}^n$ and such that if $a_{i,j} \neq 0$ for some $i \neq j$, then either $i \in S_1$ and $j \in S_2$ or $i \in S_2$ and $j \in S_1$.*

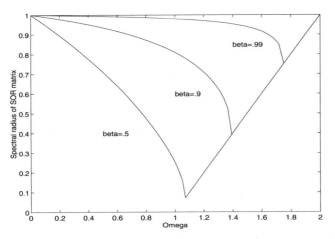

FIG. 10.1. *Spectral radius of the SOR matrix for different values of ω and $\beta = \rho(G_J)$.*

THEOREM 10.1.4. *A matrix A has Property A if and only if A is a diagonal matrix or else there exists a permutation matrix P such that $P^{-1}AP$ has the form*

(10.19)
$$\begin{pmatrix} D_1 & B \\ C & D_2 \end{pmatrix},$$

where D_1 and D_2 are square diagonal matrices.

Proof. If A has Property A, then if S_1 or S_2 is empty, A is a diagonal matrix. Otherwise, order the rows and columns of A with indices in S_1 first, followed by those with indices in S_2. From the definition of S_1 and S_2, it follows that the two diagonal blocks of order $\text{card}(S_1)$ and $\text{card}(S_2)$ will be diagonal matrices.

Conversely, if A can be permuted into the form (10.19), then take S_1 to be the set of indices corresponding to the first diagonal block and S_2 to be those corresponding to the second diagonal block. Then S_1 and S_2 satisfy the properties required in the definition of Property A.   □

We state without proof the following theorem. For a proof see, e.g., [83].

THEOREM 10.1.5. *If a matrix A has Property A then there is a permutation matrix P such that $P^{-1}AP$ satisfies (10.9).*

The Poisson Equation. We saw an example earlier of a matrix with Property A, namely, the matrix arising from a 5-point finite difference approximation to Poisson's equation on a square. By numbering the nodes of the grid in a red–black checkerboard fashion, we obtained a matrix of the form (10.19). It turns out that even if the natural ordering of nodes is used, the assumption (10.9) is satisfied for this matrix.

The eigenvalues of this matrix are known explicitly and are given in Theorem 9.1.2. If we assume that $h_x = h_y \equiv h$ and scale the matrix to have ones on its diagonal, then these eigenvalues are

$$\lambda_{i,k} = \sin^2\left(\frac{i\pi}{2(m+1)}\right) + \sin^2\left(\frac{k\pi}{2(m+1)}\right), \quad i,k = 1,\ldots,m,$$

where $m = n_x = n_y$. The eigenvalues of the Jacobi iteration matrix G_J are one minus these values, so we have

$$
\begin{aligned}
\rho(G_J) &= \max_{i,k}\left|1 - \sin^2\left(\frac{i\pi}{2(m+1)}\right) - \sin^2\left(\frac{k\pi}{2(m+1)}\right)\right| \\
(10.20) &= 1 - \frac{\pi^2}{2}h^2 + O(h^4),
\end{aligned}
$$

where the last equality comes from setting $i = k = 1$ or $i = k = m$ to obtain the maximum absolute value and then using a Taylor expansion for $\sin(x)$.

Knowing the value of $\rho(G_J)$, Theorem 10.1.3 tells us the optimal value of ω as well as the convergence rate of the SOR iteration for this and other values of ω. It follows from Theorem 10.1.3 that

$$\omega_{opt} = \frac{2}{1 + \sqrt{\pi^2 h^2 + O(h^4)}} = 2(1 - \pi h) + O(h^2)$$

and, therefore,

(10.21) $$\rho(G_{\omega_{opt}}) = 1 - 2\pi h + O(h^2).$$

In contrast, for $\omega = 1$, Theorem 10.1.3 shows that the spectral radius of the Gauss–Seidel iteration matrix is

(10.22) $$\rho(G_1) = 1 - \pi^2 h^2 + O(h^4).$$

Comparing (10.20–10.22) and ignoring higher order terms in h, it can be seen that while the asymptotic convergence rate of the Gauss–Seidel method is twice that of Jacobi's method, the difference between the Gauss–Seidel method and SOR with the optimal ω is much greater. Looking at the reduction in the log of the error for each method, we see that while the log of the error at consecutive steps differs by $O(h^2)$ for the Jacobi and Gauss–Seidel methods, it differs by $O(h)$ for SOR with the optimal ω.

Figure 10.2 shows a plot of the convergence of these three methods as well as the unpreconditioned CG algorithm for $h = 1/51$. A random solution was set and the right-hand side was computed. The 2-norm of the error is plotted. While the SOR method is a great improvement over the Jacobi and Gauss–Seidel iterations, we see that even for this moderate value of h all of the methods require many iterations to obtain a good approximate solution. As already noted, the CG iteration is more appropriate than simple iteration for symmetric positive definite problems such as this, and the remaining chapters of this book will discuss preconditioners designed to further enhance the convergence rate.

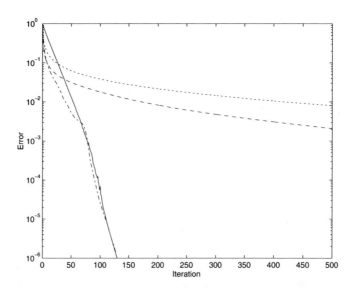

FIG. 10.2. *Convergence of iterative methods for the model problem, $h = 1/51$. Jacobi (dotted), Gauss–Seidel (dashed), SOR with optimal ω (solid), unpreconditioned CG (dash–dot).*

10.2. The Perron–Frobenius Theorem.

A powerful theory is available for comparing asymptotic convergence rates of simple iteration methods when used with a class of splittings known as "regular splittings." This theory is based on the work of Perron and Frobenius on nonnegative matrices. The Perron–Frobenius theorem is an important tool in many areas of applied linear algebra. We include here proofs of only parts of that theory. For a more complete exposition, see [80], from which this material was taken.

Notation. We will use the notation $A \geq B$ $(A > B)$ to mean that each entry of the real matrix A is greater than or equal to (strictly greater than) the corresponding entry of B. The matrix with (i, j)-entry $|a_{ij}|$ will be denoted by $|A|$. The matrix A is called *positive* (*nonnegative*) if $A > 0$ $(A \geq 0)$.

Let A and B be n-by-n matrices and let v be an n-vector. It is left as an exercise to show the following results:

10.2a. $|A^k| \leq |A|^k$ for all $k = 1, 2, \ldots$.

10.2b. If $0 \leq A \leq B$, then $0 \leq A^k \leq B^k$ for all $k = 1, 2, \ldots$.

10.2c. If $A > 0$, then $A^k > 0$ for all $k = 1, 2, \ldots$.

10.2d. If $A > 0$ and $v \geq 0$ and v is not the 0 vector, then $Av > 0$.

10.2e. If $A \geq 0$ and $v \geq 0$ and $Av \geq \alpha v$ for some $\alpha > 0$, then $A^k v \geq \alpha^k v$ for all $k = 1, 2, \ldots$.

THEOREM 10.2.1. *Let A and B be n-by-n matrices. If $|A| \leq B$, then $\rho(A) \leq \rho(|A|) \leq \rho(B)$.*

Proof. It follows from exercises 10.2a and 10.2b that for every $k = 1, 2, \ldots$, we have $|A^k| \leq |A|^k \leq B^k$, so the Frobenius norms of these matrices satisfy

$$(10.23) \qquad \|A^k\|_F^{1/k} \leq \| \, |A|^k \, \|_F^{1/k} \leq \|B^k\|_F^{1/k}.$$

Since the spectral radius of a matrix C is just $\lim_{k \to \infty} |||C^k|||^{1/k}$, where $||| \cdot |||$ is any matrix norm (Corollary 1.3.1), taking limits in (10.23) gives $\rho(A) \leq \rho(|A|) \leq \rho(B)$. □

COROLLARY 10.2.1. *Let A and B be n-by-n matrices. If $0 \leq A \leq B$, then $\rho(A) \leq \rho(B)$.*

COROLLARY 10.2.2. *Let A and B be n-by-n matrices. If $0 \leq A < B$, then $\rho(A) < \rho(B)$.*

Proof. There is a number $\alpha > 1$ such that $0 \leq A \leq \alpha A < B$. It follows from Corollary 10.2.1 that $\rho(B) \geq \alpha\rho(A)$, so if $\rho(A) \neq 0$, then $\rho(B) > \rho(A)$. If $\rho(A) = 0$, consider the matrix C with $(1,1)$-entry equal to $b_{11} > 0$ and all other entries equal to zero. The spectral radius of this matrix is b_{11}, and we have $C = |C| \leq B$, so $\rho(B) \geq b_{11} > 0$. □

In 1907, Perron proved important results for *positive* matrices. Some of these results are contained in the following theorem.

THEOREM 10.2.2. *Let A be an n-by-n matrix and suppose $A > 0$. Then $\rho(A) > 0$, $\rho(A)$ is an eigenvalue of A, and there is a positive vector v such that $Av = \rho(A)v$.*

Proof. It follows from Corollary 10.2.2 that $\rho(A) > 0$. By definition of the spectral radius, there is an eigenvalue λ with $|\lambda| = \rho(A)$. Let v be an associated nonzero eigenvector. We have

$$\rho(A)|v| = |\lambda| \cdot |v| = |\lambda v| = |Av| \leq |A| \cdot |v| = A|v|,$$

so $y \equiv A|v| - \rho(A)|v| \geq 0$. If y is the 0 vector, then this implies that $\rho(A)$ is an eigenvalue of A with the nonnegative eigenvector $|v|$. If $|v|$ had a zero component, then that component of $A|v|$ would have to be zero, and since each entry of A is positive, this would imply that v is the 0 vector (Exercise 10.2d), which is a contradiction. Thus, if y is the 0 vector, Theorem 10.2.2 is proved.

If y is not the 0 vector, then $Ay > 0$ (Exercise 10.2d); setting $z \equiv A|v| > 0$, we have $0 < Ay = Az - \rho(A)z$ or $Az > \rho(A)z$. It follows that there is some number $\alpha > \rho(A)$ such that $Az \geq \alpha z$. From Exercise 10.2e, it follows that for every $k \geq 1$, $A^k z \geq \alpha^k z$. From this we conclude that $\|A^k\|^{1/k} \geq \alpha > \rho(A)$ for all k. But since $\lim_{k \to \infty} \|A^k\|^{1/k} = \rho(A)$, this leads to the contradiction $\rho(A) \geq \alpha > \rho(A)$. □

Theorem 10.2.2 is part of the Perron theorem, which also states that there is a unique eigenvalue λ with modulus equal to $\rho(A)$ and that this eigenvalue is simple.

THEOREM 10.2.3 (Perron). *If A is an n-by-n matrix and $A > 0$, then*

(a) $\rho(A) > 0$;

(b) $\rho(A)$ *is a simple eigenvalue of* A;

(c) $\rho(A)$ *is the unique eigenvalue of maximum modulus; that is, for any other eigenvalue* λ *of* A, $|\lambda| < \rho(A)$; *and*

(d) *there is a vector* v *with* $v > 0$ *such that* $Av = \rho(A)v$.

The unique normalized eigenvector characterized in Theorem 10.2.3 is often called the *Perron vector* of A; $\rho(A)$ is often called the *Perron root* of A.

In many instances we will be concerned with *nonnegative* matrices that are not necessarily positive, so it is desirable to extend the results of Perron to this case. Some of the results can be extended just by taking suitable limits, but, unfortunately, limit arguments are only partially applicable. The results of Perron's theorem that generalize by taking limits are contained in the following theorem.

THEOREM 10.2.4. *If* A *is an n-by-n matrix and* $A \geq 0$, *then* $\rho(A)$ *is an eigenvalue of* A *and there is a nonnegative vector* $v \geq 0$, *with* $\|v\| = 1$, *such that* $Av = \rho(A)v$.

Proof. For any $\epsilon > 0$, define $A(\epsilon) \equiv [a_{ij} + \epsilon] > 0$. Let $v(\epsilon) > 0$ with $\|v(\epsilon)\| = 1$ denote the Perron vector of $A(\epsilon)$ and $\rho(\epsilon)$ the Perron root. Since the set of vectors $v(\epsilon)$ is contained in the compact set $\{w : \|w\| = 1\}$, there is a monotone decreasing sequence $\epsilon_1 > \epsilon_2 > \ldots$ with $\lim_{k \to \infty} \epsilon_k = 0$ such that $\lim_{k \to \infty} v(\epsilon_k) \equiv v$ exists and satisfies $\|v\| = 1$. Since $v(\epsilon_k) > 0$, it follows that $v \geq 0$.

By Theorem 10.2.1, the sequence of numbers $\{\rho(\epsilon_k)\}_{k=1,2,\ldots}$ is a monotone decreasing sequence. Hence $\rho \equiv \lim_{k \to \infty} \rho(\epsilon_k)$ exists and $\rho \geq \rho(A)$. But from the fact that

$$\begin{aligned} Av &= \lim_{k \to \infty} A(\epsilon_k)v(\epsilon_k) = \lim_{k \to \infty} \rho(\epsilon_k)v(\epsilon_k) \\ &= \lim_{k \to \infty} \rho(\epsilon_k) \lim_{k \to \infty} v(\epsilon_k) = \rho v \end{aligned}$$

and the fact that v is not the zero vector, it follows that ρ is an eigenvalue of A and so $\rho \leq \rho(A)$. Hence it must be that $\rho = \rho(A)$. $\square$

The parts of Theorem 10.2.3 that are not contained in Theorem 10.2.4 do not carry over to all nonnegative matrices. They can, however, be extended to *irreducible* nonnegative matrices, and this extension was carried out by Frobenius.

DEFINITION 10.2.1. *Let* A *be an n-by-n matrix. The* graph *of* A *is*

(10.24) $$G(A) = \{(i,j) : a_{ij} \neq 0\}.$$

The set $G(A)$ can be visualized as follows. For each integer $i = 1, \ldots, n$, draw a vertex, and for each pair $(i,j) \in G(A)$, draw a directed edge from vertex i to vertex j. This is illustrated in Figure 10.3.

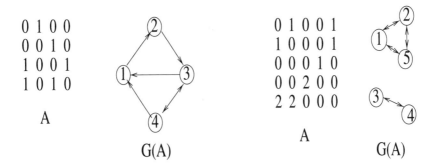

FIG. 10.3.   *Graph of a matrix.*

DEFINITION 10.2.2. *An n-by-n matrix A is called* irreducible *if every vertex in the graph of A is connected to every other vertex through a chain of edges. Otherwise, A is called* reducible.

The matrix A is reducible if and only if there is an ordering of the indices such that A takes the form

(10.25)
$$\begin{pmatrix} A_{11} & A_{12} \\ 0 & A_{22} \end{pmatrix},$$

where A_{11} and A_{22} are square blocks of dimension greater than or equal to 1. To see this, first suppose that A is of the form (10.25) for some ordering of the indices. Let I_1 be the set of row numbers of the entries of A_{11} and let I_2 be the set of row numbers of the entries of A_{22}. If $j \in I_2$ is connected to $i \in I_1$, then somewhere in the path from j to i there must be an edge connecting an element of I_2 to an element of I_1, but this would correspond to a nonzero entry in the $(2,1)$ block of (10.25). Conversely, if A is reducible, then there must be indices j and i such that j is not connected to i. Let $I_1 = \{k : k \text{ is connected to } i\}$ and let I_2 consist of the remaining indices. The sets I_1 and I_2 are nonempty, since $i \in I_1$ and $j \in I_2$. Enumerate first I_1, then I_2. If an entry in I_2 were connected to any entry in I_1, it would be connected to i, which is a contradiction. Therefore, the $(2,1)$ block in the representation of A using this ordering would have to be 0, as in (10.25). The matrix on the left in Figure 10.3 is irreducible, while that on the right is reducible.

THEOREM 10.2.5 (Perron–Frobenius). *Let A be an n-by-n real matrix and suppose that A is irreducible and nonnegative. Then*

(a) $\rho(A) > 0$;

(b) $\rho(A)$ *is a simple eigenvalue of A;*

(c) *if A has exactly k eigenvalues of maximum modulus $\rho(A)$, then these eigenvalues are the kth roots of unity times $\rho(A)$: $\lambda_j = e^{2\pi i j/k}\rho(A)$; and*

(d) *there is a vector v with $v > 0$ such that $Av = \rho(A)v$.*

10.3. Comparison of Regular Splittings.

We now use the Perron–Frobenius theorem to compare "regular splittings" when the coefficient matrix A is "inverse-positive." The main results of this section (Theorem 10.3.1 and Corollaries) are due to Varga [135].

DEFINITION 10.3.1. *For n-by-n real matrices A, M, and N, the splitting $A = M - N$ is a regular splitting if M is nonsingular with $M^{-1} \geq 0$ and $M \geq A$.*

THEOREM 10.3.1. *Let $A = M - N$ be a regular splitting of A, where $A^{-1} \geq 0$. Then*

$$\rho(M^{-1}N) = \frac{\rho(A^{-1}N)}{1 + \rho(A^{-1}N)} < 1.$$

Proof. Since $M^{-1}A = I - M^{-1}N$ is nonsingular, it follows that $M^{-1}N$ cannot have an eigenvalue equal to 1. Since $M^{-1}N \geq 0$, this, combined with Theorem 10.2.4, shows that $\rho(M^{-1}N)$ cannot be 1. It also follows from Theorem 10.2.4 that there is a vector $v \geq 0$ such that $M^{-1}Nv = \rho(M^{-1}N)v$. Now we can also write

$$A^{-1}N = (I - M^{-1}N)^{-1}M^{-1}N,$$

so

(10.26)
$$A^{-1}Nv = \frac{\rho(M^{-1}N)}{1 - \rho(M^{-1}N)}v.$$

If $\rho(M^{-1}N) > 1$, then this would imply that $A^{-1}Nv$ has negative components, which is impossible since $A^{-1} \geq 0$, $N \geq 0$, and $v \geq 0$. This proves that $\rho(M^{-1}N) < 1$. It also follows from (10.26) that $\rho(M^{-1}N)/(1 - \rho(M^{-1}N))$ is an eigenvalue of $A^{-1}N$, so we have

$$\rho(A^{-1}N) \geq \frac{\rho(M^{-1}N)}{1 - \rho(M^{-1}N)}$$

or, equivalently, since $1 - \rho(M^{-1}N) > 0$,

(10.27)
$$\rho(M^{-1}N) \leq \frac{\rho(A^{-1}N)}{1 + \rho(A^{-1}N)}.$$

Now, we also have $A^{-1}N \geq 0$, from which it follows by Theorem 10.2.4 that there is a vector $w \geq 0$ such that $A^{-1}Nw = \rho(A^{-1}N)w$. Using the relation

$$M^{-1}N = (A + N)^{-1}N = (I + A^{-1}N)^{-1}A^{-1}N,$$

we can write

$$M^{-1}Nw = \frac{\rho(A^{-1}N)}{1 + \rho(A^{-1}N)}w,$$

so $\rho(A^{-1}N)/(1 + \rho(A^{-1}N))$ is an eigenvalue of $M^{-1}N$. It follows that

$$\rho(M^{-1}N) \geq \frac{\rho(A^{-1}N)}{1 + \rho(A^{-1}N)},$$

and combining this with (10.27), the theorem is proved.   □

From Theorem 10.3.1 and the fact that $x/(1+x)$ is an increasing function of x, the following corollary is obtained.

COROLLARY 10.3.1. *Let $A = M_1 - N_1 = M_2 - N_2$ be two regular splittings of A, where $A^{-1} \geq 0$. If $N_1 \leq N_2$ then*

$$\rho(M_1^{-1}N_1) \leq \rho(M_2^{-1}N_2).$$

With the slightly stronger assumption that $A^{-1} > 0$, the inequalities in Corollary 10.3.1 can be replaced by strict inequalities.

COROLLARY 10.3.2. *Let $A = M_1 - N_1 = M_2 - N_2$ be two regular splittings of A, where $A^{-1} > 0$. If $N_1 \leq N_2$ and neither N_1 nor $N_2 - N_1$ is the null matrix, then*

$$0 < \rho(M_1^{-1}N_1) < \rho(M_2^{-1}N_2) < 1.$$

It may not be easy to determine if the inverse of a coefficient matrix A is nonnegative or positive, which are the conditions required in Corollaries 10.3.1 and 10.3.2. In [81, pp. 114–115], a number of equivalent criteria are established. We state a few of these here.

DEFINITION 10.3.2. *An n-by-n matrix A is called an M-matrix if*

(i) $a_{ii} > 0$, $i = 1, \ldots, n$,

(ii) $a_{ij} \leq 0$, $i, j = 1, \ldots, n$, $j \neq i$, *and*

(iii) *A is nonsingular and $A^{-1} \geq 0$.*

The name "M-matrix" was introduced by Ostrowski in 1937 as an abbreviation for "Minkowskische Determinante."

THEOREM 10.3.2 (see [81]). *Let A be a real n-by-n matrix with nonpositive off-diagonal entries. The following statements are equivalent:*

1. *A is an M-matrix.*

2. *A is nonsingular and $A^{-1} \geq 0$. (Note that condition (i) in the definition of an M-matrix is not necessary. It is implied by the other two conditions.)*

3. *All eigenvalues of A have positive real part. (A matrix A with this property is called* positive stable, *whether or not its off-diagonal entries are nonpositive.)*

4. *Every real eigenvalue of A is positive.*

5. *All principal minors of A are M-matrices.*

6. *A can be factored in the form $A = LU$, where L is lower triangular, U is upper triangular, and all diagonal entries of each are positive.*

7. *The diagonal entries of A are positive, and AD is strictly row diagonally dominant for some positive diagonal matrix D.*

It was noted in section 9.2 that under assumption (9.35) the coefficient matrix (9.31) arising from the transport equation has positive diagonal entries and nonpositive off-diagonal entries. It was also noted that the matrix is weakly row diagonally dominant. It is only weakly diagonally dominant because off-diagonal entries of the rows in the last block sum to 1. If we assume, however, that $\gamma \equiv \sup_x \sigma_s(x)/\sigma_t(x) < 1$, then the other rows are strongly diagonally dominant. If the last block column is multiplied by a number greater than 1 but less than γ^{-1}, then the resulting matrix will be strictly row diagonally dominant. Thus this matrix satisfies criterion (7) of Theorem 10.3.2, and therefore it is an M-matrix. The block Gauss–Seidel splitting described in section 9.2 is a regular splitting, so by Theorem 10.3.1, iteration (9.34) converges. Additionally, if the initial error has all components of one sign, then the same holds for the error at each successive step, since the iteration matrix $I - M^{-1}A \equiv M^{-1}N$ has nonnegative entries. This property is often important when subsequent computations with the approximate solution vector expect a nonnegative vector because the physical flux is nonnegative.

In the case of real symmetric matrices, criterion (3) (or (4)) of Theorem 10.3.2 implies that a positive definite matrix with nonpositive off-diagonal entries is an M-matrix. We provide a proof of this part.

DEFINITION 10.3.3. *A real matrix A is a* Stieltjes matrix *if A is symmetric positive definite and the off-diagonal entries of A are nonpositive.*

THEOREM 10.3.3. *Any Stieltjes matrix is an M-matrix.*

Proof. Let A be a Stieltjes matrix. The diagonal elements of A are positive because A is positive definite, so we need only verify that $A^{-1} \geq 0$. Write $A = D - C$, where $D = \text{diag}(A)$ is positive and C is nonnegative. Since A is positive definite, it is nonsingular, and $A^{-1} = [D(I - B)]^{-1} = (I - B)^{-1}D^{-1}$, where $B = D^{-1}C$. If $\rho(B) < 1$, then the inverse of $I - B$ is given by the Neumann series

$$(I - B)^{-1} = I + B + B^2 + \cdots,$$

and since $B \geq 0$ it would follow that $(I - B)^{-1} \geq 0$ and, hence, $A^{-1} \geq 0$. Thus, we need only show that $\rho(B) < 1$.

Suppose $\rho(B) \geq 1$. Since $B \geq 0$, it follows from Theorem 10.2.4 that $\rho(B)$ is an eigenvalue of B. But then $D^{-1}A = I - B$ must have a nonpositive eigenvalue, $1 - \rho(B)$. This matrix is similar to the symmetric positive definite matrix $D^{-1/2}AD^{-1/2}$, so we have a contradiction. Thus $\rho(B) < 1$.   □

The matrix arising from the diffusion equation defined in (9.6–9.8) is a Stieltjes matrix and, hence, an M-matrix.

It follows from Corollary 10.3.1 that if A is an M-matrix then the asymptotic convergence rate of the Gauss–Seidel iteration is at least as good as that of Jacobi's method. In this case, both methods employ regular splittings, and the lower triangle of A, used in the Gauss–Seidel iteration, is closer (elementwise) to A than the diagonal of A used in the Jacobi iteration. If the matrix A is also inverse positive, $A^{-1} > 0$, then Corollary 10.3.2 implies that the asymptotic convergence rate of the Gauss–Seidel iteration is strictly better than that of Jacobi's method. (A stronger relation was proved in Corollary 10.1.1, but this was only for matrices satisfying (10.9).) Among all diagonal matrices M whose diagonal entries are greater than or equal to those of A, however, Corollary 10.3.2 implies that the Jacobi splitting $M = \text{diag}(A)$ is the best. Similarly, when considering regular splittings in which the matrix M is restricted to have a certain sparsity pattern (e.g., banded with a fixed bandwidth), Corollary 10.3.2 implies that the best choice of M, as far as asymptotic convergence rate of the simple iteration method is concerned, is to take the variable entries of M to be equal to the corresponding entries of A.

Corollaries 10.3.1 and 10.3.2 confirm one's intuition, in the special case of regular splittings of inverse-nonnegative or inverse-positive matrices, that the closer the preconditioner M is to the coefficient matrix A, the better the convergence of the preconditioned simple iteration (at least asymptotically). Of course, many regular splittings cannot be compared using these theorems because certain entries of one splitting are closer to those of A while different entries of the other are closer. Also, many of the best splittings are *not* regular splittings, so these theorems do not apply. The SOR splitting is not a regular splitting for an M-matrix if $\omega > 1$.

10.4. Regular Splittings Used with the CG Algorithm.

For Hermitian positive definite systems, the A-norm of the error in the PCG algorithm (which is the $L^{-1}AL^{-H}$-norm of the error for the modified linear system (8.2)) and the 2-norm of L^{-1} times the residual in the PMINRES algorithm can be bounded in terms of the square root of the condition number of the preconditioned matrix using (3.8) and (3.12). Hence, in measuring the effect of a preconditioner, we will be concerned not with the spectral radius of $I - M^{-1}A$ but with the condition number of $L^{-1}AL^{-H}$ (or, equivalently, with the ratio of largest to smallest eigenvalue of $M^{-1}A$).

With slight modifications, Corollaries 10.3.1 and 10.3.2 can also be used to compare condition numbers of PCG or PMINRES iteration matrices when A, M_1, and M_2 are real symmetric and positive definite. Note that some modifications will be required, however, because unlike simple iteration, the PCG and PMINRES algorithms are insensitive to scalar multiples in the preconditioner; that is, the approximations generated by these algorithms with preconditioner M are the same as those generated with preconditioner cM for any $c > 0$ (Exercise 10.3).

THEOREM 10.4.1. *Let A, M_1, and M_2 be symmetric, positive definite* *matrices satisfying the hypotheses of Corollary 10.3.1, and suppose that the largest eigenvalue of $M_2^{-1}A$ is greater than or equal to 1. Then the ratios of largest to smallest eigenvalues of $M_1^{-1}A$ and $M_2^{-1}A$ satisfy*

(10.28)
$$\frac{\lambda_{max}(M_1^{-1}A)}{\lambda_{min}(M_1^{-1}A)} < 2\,\frac{\lambda_{max}(M_2^{-1}A)}{\lambda_{min}(M_2^{-1}A)}.$$

Proof. Since the elements of $M_2^{-1}N_2$ are nonnegative, it follows from Theorem 10.2.4 that its spectral radius is equal to its (algebraically) largest eigenvalue:

$$\rho(M_2^{-1}N_2) = \rho(I - M_2^{-1}A) = 1 - \lambda_{min}(M_2^{-1}A).$$

The result $\rho(M_1^{-1}N_1) \le \rho(M_2^{-1}N_2)$ from Corollary 10.3.1 implies that

$$1 - \lambda_{min}(M_1^{-1}A) \le 1 - \lambda_{min}(M_2^{-1}A) \quad \text{and} \quad \lambda_{max}(M_1^{-1}A) - 1 \le 1 - \lambda_{min}(M_2^{-1}A)$$

or, equivalently,

$$\lambda_{min}(M_1^{-1}A) \ge \lambda_{min}(M_2^{-1}A) \quad \text{and} \quad \lambda_{max}(M_1^{-1}A) \le 2 - \lambda_{min}(M_2^{-1}A).$$

Dividing the second inequality by the first gives

$$\frac{\lambda_{max}(M_1^{-1}A)}{\lambda_{min}(M_1^{-1}A)} \le \frac{\lambda_{max}(M_2^{-1}A)}{\lambda_{min}(M_2^{-1}A)}\left(\frac{2 - \lambda_{min}(M_2^{-1}A)}{\lambda_{max}(M_2^{-1}A)}\right).$$

Since, by assumption, $\lambda_{max}(M_2^{-1}A) \ge 1$ and since $\rho(M_2^{-1}N_2) < 1$ implies that $\lambda_{min}(M_2^{-1}A) > 0$, the second factor on the right-hand side is less than 2, and the theorem is proved. $\square$

THEOREM 10.4.2. *The assumption in Theorem 10.4.1 that the largest eigenvalue of $M_2^{-1}A$ is greater than or equal to 1 is satisfied if A and M_2 have at least one diagonal element in common.*

Proof. If A and M_2 have a diagonal element in common, then the symmetric matrix N_2 has a zero diagonal element. This implies that $M_2^{-1}N_2$ has a nonpositive eigenvalue since the smallest eigenvalue of this matrix satisfies

$$\min_{v \ne 0} \frac{v^*N_2v}{v^*M_2v} \le \frac{\xi_j^*N_2\xi_j}{\xi_j^*M_2\xi_j} = 0$$

if ξ_j is the vector with a 1 in the position of this zero diagonal element and 0's elsewhere. Therefore, $M_2^{-1}A = I - M_2^{-1}N_2$ has an eigenvalue greater than or equal to 1. $\square$

Theorems 10.4.1 and 10.4.2 show that once a pair of regular splittings have been scaled properly for comparison (that is, M_2 has been multiplied by a constant, if necessary, so that A and M_2 have at least one diagonal element in common), the one that is closer to A elementwise gives a smaller condition number for the PCG or PMINRES iteration matrix (except possibly for a factor of 2). This means that the Chebyshev bound (3.8) on the error at each step will be smaller (or, at worst, only slightly larger) for the closer preconditioner. Other properties, however, such as tight clustering of most of the eigenvalues, also affect the convergence rate of PCG and PMINRES. Unfortunately, it would be difficult to provide general comparison theorems based on all of these factors, so the condition number is generally used for this purpose.

10.5. Optimal Diagonal and Block Diagonal Preconditioners.

Aside from regular splittings, about the only class of preconditioners among which an optimal or near optimal preconditioner is known is the class of diagonal or block-diagonal preconditioners. If "optimality" is defined in terms of the symmetrically preconditioned matrix having a small condition number, then the (block) diagonal of a Hermitian positive definite matrix A is close to the best (block) diagonal preconditioner.

Recall the definition of Property A from section 10.1. A matrix with this property is also said to be 2-*cyclic*. Moreover, we can make the following more general definition.

DEFINITION 10.5.1. *A matrix A is* block 2-cyclic *if it can be permuted into the form*

$$A = \begin{pmatrix} D_1 & B \\ C & D_2 \end{pmatrix},$$

where D_1 and D_2 are block diagonal matrices

$$D_i = \begin{pmatrix} D_{i,1} & & \\ & \ddots & \\ & & D_{i,m_i} \end{pmatrix}, \quad i = 1,2, \quad D_{i,j} \in \mathbf{C}^{n_{i,j} \times n_{i,j}}.$$

Forsythe and Strauss [52] showed that for a Hermitian positive definite matrix A in 2-cyclic form, the optimal diagonal preconditioner is $M = \mathrm{diag}(A)$. Eisenstat, Lewis, and Schultz [41] later generalized this to cover matrices in block 2-cyclic form with block diagonal preconditioners. They showed that if each block $D_{i,j}$ is the identity, then A is optimally scaled with respect to all block diagonal matrices with blocks of order $n_{i,j}$. The following slightly stronger result is due to Elsner [42].

THEOREM 10.5.1 (Elsner). *If a Hermitian positive definite matrix A has the form*

(10.29)
$$A = \begin{pmatrix} I_{n_1} & B \\ B^H & I_{n_2} \end{pmatrix},$$

then $\kappa(A) \leq \kappa(D^H A D)$ for any nonsingular D of the form

(10.30)
$$\begin{pmatrix} D_1 & 0 \\ 0 & D_2 \end{pmatrix}, \quad D_1 \in \mathbf{C}^{n_1 \times n_1}, \quad D_2 \in \mathbf{C}^{n_2 \times n_2}.$$

Proof. If λ is an eigenvalue of A with an eigenvector whose first block is v and whose second block is w (which we will denote as $(v; w)$), then it follows from (10.29) that

$$Bw = (\lambda - 1)v, \quad B^H v = (\lambda - 1)w.$$

From this we conclude that $(v; -w)^T$ is an eigenvector of A with eigenvalue $2 - \lambda$, since

$$A \begin{pmatrix} v \\ -w \end{pmatrix} = \begin{pmatrix} v - Bw \\ B^H v - w \end{pmatrix} = (2 - \lambda) \begin{pmatrix} v \\ -w \end{pmatrix}.$$

It follows that if λ_n is the largest eigenvalue of A, then $\lambda_1 = 2 - \lambda_n$ is the smallest, and $\kappa(A) = \lambda_n/(2 - \lambda_n)$.

Let

$$S = \begin{pmatrix} I_{n_1} & \\ & -I_{n_2} \end{pmatrix},$$

so $S(v; w) = (v; -w)$. If $Az = \lambda_n z$, then $SAz = \lambda_n Sz$ and

$$A^{-1}SAz = \frac{\lambda_n}{2 - \lambda_n}Sz, \quad SA^{-1}SAz = \frac{\lambda_n}{2 - \lambda_n}z.$$

Thus we have

$$\rho(SA^{-1}SA) \geq \frac{\lambda_n}{2 - \lambda_n} = \kappa(A),$$

and for any nonsingular matrix D, we can write

$$(10.31) \quad \kappa(A) \leq \rho(SA^{-1}SA) = \rho(D^{-1}SA^{-1}SAD) \leq \|D^{-1}SA^{-1}SAD\|.$$

Now, if D is of the form (10.30), then S and D commute. Also, $\|S\| = 1$, so we have

$$
\begin{aligned}
\|D^{-1}SA^{-1}SAD\| &= \|S(D^{-1}A^{-1}D^{-H})S(D^H AD)\| \\
(10.32) \quad &\leq \|D^{-1}A^{-1}D^{-H}\| \cdot \|D^H AD\| = \kappa(D^H AD).
\end{aligned}
$$

Combining (10.31) and (10.32) gives the desired result. □

Suppose A is not of the form (10.29) but can be permuted into that form, say, $A = P^T \tilde{A} P$, where P is a permutation matrix and $\tilde{A}$ is of the form (10.29). Then for any block-diagonal matrix D of the form (10.30), we can write

$$\kappa(D^H AD) = \kappa(PD^H P^T \tilde{A} PDP^T).$$

If the permutation is such that PDP^T is a block-diagonal matrix of the form (10.30), then A, like $\tilde{A}$, is optimally scaled among such block-diagonal matrices; if $\kappa(D^H AD)$ were less than $\kappa(A)$ for some D of the form (10.30), then $\kappa(\tilde{D}^H \tilde{A} \tilde{D})$ would be less than $\kappa(\tilde{A})$, where $\tilde{D} = PDP^T$, which is a contradiction. In particular, if A has Property A then the optimal diagonal preconditioner is $M = \text{diag}(A)$, since if D is diagonal then $P^T DP$ is diagonal for any permutation matrix P. If A is written in a block form, where the blocks can be permuted into block 2-cyclic form (without permuting entries from one block to another), then the optimal block diagonal preconditioner (with the same size blocks) is $M = \text{block diag}(A)$.

Theorem 10.5.1 implies that for Hermitian positive definite block 2-cyclic matrices, the block diagonal of the matrix is the best block-diagonal preconditioner (in terms of minimizing the condition number of the preconditioned matrix). For arbitrary Hermitian positive definite matrices, the block diagonal of the matrix is almost optimal. The following theorem of van der Sluis [131] deals with ordinary diagonal preconditioners, while the next theorem, due to Demmel [31] (and stated here without proof), deals with the block case.

THEOREM 10.5.2 (van der Sluis). *If a Hermitian positive definite matrix A has all diagonal elements equal, then*

$$(10.33) \qquad \kappa(A) \leq m \cdot \min_{D \in \mathcal{D}} \kappa(DAD),$$

where $\mathcal{D} = \{positive\ definite\ diagonal\ matrices\}$ and m is the maximum number of nonzeros in any row of A.

Proof. Write $A = U^H U$, where U is upper triangular. Since A has equal diagonal elements, say, 1, each column of U has norm 1. Also, each off-diagonal entry of A has absolute value less than or equal to 1, since $|a_{ij}| = |u_i^H u_j| \leq \|u_i\| \cdot \|u_j\| \leq 1$, where u_i and u_j are the ith and jth columns of U. Additionally, it follows from Gerschgorin's theorem that

$$\|A\| = \lambda_n(A) \leq \max_i \sum_j |a_{ij}| \leq m.$$

For any nonsingular matrix D we can write

$$\begin{aligned}
\kappa(D^H AD) &= \|D^H AD\| \cdot \|D^{-1} A^{-1} D^{-H}\| \\
&= \|D^H U^H UD\| \cdot \|D^{-1} U^{-1} U^{-H} D^{-H}\| \\
&= \|UD\|^2 \cdot \|D^{-1} U^{-1}\|^2.
\end{aligned}$$

Now, $\|D^{-1} U^{-1}\|^2 \geq \|U^{-1}\|^2 / \|D\|^2 = \|A^{-1}\| / \|D\|^2$, so we have

$$(10.34) \quad \kappa(D^H AD) \geq \|UD\|^2 \frac{\|A^{-1}\|}{\|D\|^2} = \kappa(A) \cdot \frac{\|UD\|^2}{\|A\| \cdot \|D\|^2} \geq \frac{\kappa(A)}{m} \left(\frac{\|UD\|}{\|D\|} \right)^2.$$

(Note that we have not yet made any assumption about the matrix D. The result holds for any nonsingular matrix D such that $\|UD\| \geq \|D\|$.)

Now assume that D is a positive definite diagonal matrix with largest entry d_{jj}. Let ξ_j be the jth unit vector. Then

$$(10.35) \qquad \|UD\| = \max_{\|v\|=1} \|UDv\| \geq \|UD\xi_j\| = \|d_{jj} u_j\| = \|D\|.$$

Combining (10.34) and (10.35) gives the desired result.   □

THEOREM 10.5.3 (Demmel). *If a Hermitian positive definite matrix A has all diagonal blocks equal to the identity, say*

$$A = \begin{pmatrix} I_{n_1} & A_{12} & \cdots & A_{1m} \\ A_{12}^H & I_{n_2} & \cdots & A_{2m} \\ \vdots & \vdots & \ddots & \vdots \\ A_{1m}^H & A_{2m}^H & \cdots & I_{n_m} \end{pmatrix},$$

then

$$\kappa(A) \leq m \cdot \min_{D \in \mathcal{D}_B} \kappa(D^H A D),$$

where $\mathcal{D}_B = \{$*nonsingular block-diagonal matrices with blocks of order* n_1, $\ldots, n_m\}$, *and* m *is the number of diagonal blocks in* A.

As an example, consider the matrix defined in (9.6–9.8) arising from a 5-point finite difference approximation to the diffusion equation. This matrix is block tridiagonal with n_y diagonal blocks, each of order n_x. Of all block diagonal preconditioners D with blocks of order n_x, the optimal one for minimizing the condition number of the symmetrically preconditioned matrix $D^{-1/2} A D^{-1/2}$, or the ratio of largest to smallest eigenvalue of $D^{-1} A$, is

$$D = \begin{pmatrix} S_1 & & \\ & \ddots & \\ & & S_{n_y} \end{pmatrix}.$$

It follows from Theorem 10.5.3 that this matrix D is within a factor of n_y of being optimal, but it follows from Theorem 10.5.1 that D is actually optimal because the blocks of A can be permuted into block 2-cyclic form.

These theorems on block-diagonal preconditioners establish just what one might expect—the best (or almost best) block-diagonal preconditioner M has all of its block-diagonal elements equal to the corresponding elements of A. Unfortunately, such results do not hold for matrices M with other sparsity patterns. For example, suppose one considers tridiagonal preconditioners M for the matrix in (9.6–9.8) or even for the simpler 5-point approximation to the negative Laplacian. The tridiagonal part of this matrix is

$$\frac{1}{h^2} \begin{pmatrix} 4 & -1 & & & & & & \\ -1 & \ddots & & \ddots & & & & \\ & \ddots & \ddots & & -1 & & & \\ & & -1 & 4 & 0 & & & \\ & & 0 & 4 & -1 & & & \\ & & & -1 & \ddots & \ddots & & \\ & & & & \ddots & \ddots & -1 & \\ & & & & & -1 & 4 & \\ & & & & & & & \ddots \end{pmatrix}.$$

The block-diagonal part of A is a tridiagonal matrix, and it is the optimal block-diagonal preconditioner for A, but it is *not* the optimal tridiagonal preconditioner. By replacing the zeros in A between the diagonal blocks with certain nonzero entries, one can obtain a better preconditioner. (To obtain as much as a factor of 2 improvement in the condition number, however, at least some of the replacement entries must be negative, since otherwise this would be a regular splitting and Theorem 10.4.1 would apply.) The optimal tridiagonal preconditioner for the 5-point Laplacian is not known analytically.

Based on the results of this section, it is reasonable to say that one should *always* (well, almost always) use at least the diagonal of a positive definite matrix as a preconditioner with the CG or MINRES algorithm. Sometimes matrices that arise in practice have diagonal entries that vary over many orders of magnitude. For example, a finite difference or finite element matrix arising from the diffusion equation (9.1–9.2) will have widely varying diagonal entries if the diffusion coefficient $a(x, y)$ varies over orders of magnitude. The eigenvalues of the matrix will likewise vary over orders of magnitude, although a simple diagonal scaling would greatly reduce the condition number. For such problems it is extremely important to scale the matrix by its diagonal or, equivalently, to use a diagonal preconditioner (or some more sophisticated preconditioner that implicitly incorporates diagonal scaling). The extra work required for diagonal preconditioning is minimal. Of course, for the model problem for Poisson's equation, the diagonal of the matrix is a multiple of the identity, so unpreconditioned CG and diagonally scaled CG are identical. The arguments for diagonal scaling might not apply if the unscaled matrix has special properties apart from the condition number that make it especially amenable to solution by CG or MINRES.

Exercises.

10.1. Show that the SSOR preconditioner is of the form (10.6).

10.2. Prove the results in (10.2a–e).

10.3. Show that the iterates x_k generated by the PCG algorithm with preconditioner M are the same as those generated with preconditioner cM for any $c > 0$.

10.4. The *multigroup* transport equation can be written in the form

$$\Omega \cdot \nabla \psi_g + \sigma_g \psi_g - \sum_{g'=1}^{G} \int_{S^2} d\Omega' \, \sigma_{g,g'}(r, \Omega \cdot \Omega') \psi_{g'}(r, \Omega') = f_g,$$

$$g = 1, \ldots, G,$$

where $\psi_g(r, \Omega)$ is the unknown flux associated with energy group g and $\sigma_g(r, \Omega)$, $\sigma_{g,g'}(r, \Omega \cdot \Omega')$, and $f_g(r, \Omega)$ are known cross section and source terms. (Appropriate boundary conditions are also given.) A standard method for solving this set of equations is to move the terms of the sum corresponding to different energy groups to the right-hand side and solve the resulting set of equations for $\psi_1, \ldots, \psi_G$ in increasing order of index, using the most recently updated quantities on the right-hand side; that is,

$$
\begin{aligned}
(\Omega \cdot \nabla + \sigma_g)\psi_g^{(k)} - \int_{S^2} d\Omega' \, \sigma_{g,g} \psi_g^{(k)} \;=\; & f_g + \sum_{g'<g} \int d\Omega' \, \sigma_{g,g'} \psi_{g'}^{(k)} \\
& + \sum_{g'>g} \int d\Omega' \, \sigma_{g,g'} \psi_{g'}^{(k-1)}.
\end{aligned}
$$

Identify this procedure with one of the preconditioned iterative methods described in this chapter. How might it be accelerated?

Chapter 11

Incomplete Decompositions

A number of matrix decompositions were described in section 1.3. These include the LU or Cholesky decomposition as well as the QR factorization. Each of these can be used to solve a linear system $Ax = b$. If the matrix A is sparse, however, the triangular factors L and U are usually much less sparse; this is similar for the unitary and upper triangular factors Q and R. For large sparse matrices, such as those arising from the discretization of partial differential equations, it is usually impractical to compute and work with these factors.

Instead, one might obtain an approximate factorization, say, $A \approx LU$, where L and U are sparse lower and upper triangular matrices, respectively. The product $M = LU$ then could be used as a preconditioner in an iterative method for solving $Ax = b$. In this chapter we discuss a number of such *incomplete factorizations*.

11.1. Incomplete Cholesky Decomposition.

Any Hermitian positive definite matrix A can be factored in the form $A = LL^H$, where L is a lower triangular matrix. This is called the Cholesky factorization. If A is a sparse matrix, however, such as the 5-point approximation to the diffusion equation defined in (9.6–9.8), then the lower triangular factor L is usually much less sparse than A. In this case, the entire band "fills in" during Gaussian elimination, and L has nonzeros throughout a band of width n_x below the main diagonal. The amount of work to compute L is $O(n_x^2 \cdot n_x n_y) = O(n^2)$ if $n_x = n_y$ and $n = n_x n_y$. The work required to solve a linear system with coefficient matrix L is $O(n_x \cdot n_x n_y)$ or $O(n^{3/2})$.

One might obtain an approximate factorization of A by restricting the lower triangular matrix L to have a given sparsity pattern, say, the sparsity pattern of the lower triangle of A. The nonzeros of L then could be chosen so that the product LL^H would match A in the positions where A has nonzeros, although, of course, LL^H could not match A everywhere. An approximate factorization of this form is called an *incomplete Cholesky decomposition*. The matrix $M = LL^H$ then can be used as a preconditioner in an iterative method such as the PCG algorithm. To solve a linear system $Mz = r$, one first solves the lower triangular system $Ly = r$ and then solves the upper triangular system $L^H z = y$.

The same idea can also be applied to non-Hermitian matrices to obtain an approximate LU factorization. The product $M = LU$ of the incomplete LU factors then can be used as a preconditioner in a non-Hermitian matrix iteration such as GMRES, QMR, or BiCGSTAB. The idea of generating such approximate factorizations has been discussed by a number of people, the first of whom was Varga [136]. The idea became popular when it was used by Meijerink and van der Vorst [99] to generate preconditioners for the CG method and related iterations. It has proved a very successful technique in a range of applications and is now widely used in large physics codes. The main results of this section are from [99].

We will show that the incomplete LU decomposition exists if the coefficient matrix A is an M-matrix. This result was generalized by Manteuffel [95] to cover H-matrices with positive diagonal elements. The matrix $A = [a_{ij}]$ is an H-matrix if its *comparison matrix*—the matrix with diagonal entries $|a_{ii}|$, $i = 1, \ldots, n$ and off-diagonal entries $-|a_{ij}|$, $i, j = 1, \ldots, n$, $j \neq i$—is an M-matrix. Any diagonally dominant matrix is an H-matrix, regardless of the signs of its entries.

In fact, this decomposition often exists even when A is not an H-matrix. It is frequently applied to problems in which the coefficient matrix is not an H-matrix, and entries are modified, when necessary, to make the decomposition stable [87, 95].

The proof will use two results about M-matrices, one due to Fan [47] and one due to Varga [135].

LEMMA 11.1.1 (Fan). *If $A = [a_{ij}]$ is an M-matrix, then $A^{(1)} = [a_{ij}^{(1)}]$ is an M-matrix, where $A^{(1)}$ is the matrix that arises by eliminating the first column of A using the first row.*

LEMMA 11.1.2 (Varga). *If $A = [a_{ij}]$ is an M-matrix and the elements of $B = [b_{ij}]$ satisfy*

$$0 < a_{ii} \leq b_{ii}, \quad a_{ij} \leq b_{ij} \leq 0 \text{ for } i \neq j,$$

then B is also an M-matrix.

Proof. Write $B = D - C = D(I - G)$, where $G = D^{-1}C \geq 0$. We have $B^{-1} = (I - G)^{-1}D^{-1}$, and if $\rho(G) < 1$, then

$$(I - G)^{-1} = I + G + G^2 + \cdots \geq 0,$$

so it will follow that $B^{-1} \geq 0$ and, therefore, that B is an M-matrix. To see that $\rho(G) < 1$, note that if A is written in the form $A = M - N$, where $M = \text{diag}(A)$, then this is a regular splitting, so we have $\rho(M^{-1}N) < 1$. From the assumptions on B, however, it follows that $0 \leq G \leq M^{-1}N$, so from the Perron–Frobenius theorem we have

$$\rho(G) \leq \rho(M^{-1}N) < 1. \quad \square$$

Lemma 11.1.2 also could be derived from (7) in Theorem 10.3.3.

Let P be a subset of the indices $\{(i,j) : j \neq i,\ i,j = 1,\dots,n\}$. The indices in the set P will be the ones forced to be 0 in our incomplete LU factorization. The following theorem not only establishes the existence of the incomplete LU factorization but also shows how to compute it.

THEOREM 11.1.1 (Meijerink and van der Vorst). *If $A = [a_{ij}]$ is an n-by-n M-matrix, then for every subset P of off-diagonal indices there exists a lower triangular matrix $L = [l_{ij}]$ with unit diagonal and an upper triangular matrix $U = [u_{ij}]$ such that $A = LU - R$, where*

$$l_{ij} = 0 \text{ if } (i,j) \in P, \quad u_{ij} = 0 \text{ if } (i,j) \in P, \quad \text{and} \quad r_{ij} = 0 \text{ if } (i,j) \notin P.$$

The factors L and U are unique, and the splitting $A = LU - R$ is a regular splitting.

Proof. The proof proceeds by construction through $n-1$ stages analogous to the stages of Gaussian elimination. At the kth stage, first replace the entries in the current coefficient matrix with indices (k,j) and $(i,k) \in P$ by 0. Then perform a Gaussian elimination step in the usual way: eliminate the entries in rows $k+1$ through n of column k by adding appropriate multiples of row k to rows $k+1$ through n. To make this precise, define the matrices

$$A^{(k)} \equiv \left[a_{ij}^{(k)}\right], \quad \tilde{A}^{(k)} \equiv \left[\tilde{a}_{ij}^{(k)}\right], \quad L^{(k)} \equiv \left[l_{ij}^{(k)}\right], \quad R^{(k)} \equiv \left[r_{ij}^{(k)}\right]$$

by the relations

$$A^{(0)} = A, \quad \tilde{A}^{(k)} = A^{(k-1)} + R^{(k)}, \quad A^{(k)} = L^{(k)}\tilde{A}^{(k)}, \quad k = 1,\dots,n-1,$$

where $R^{(k)}$ is zero except in positions $(k,j) \in P$ and in positions $(i,k) \in P$, where $r_{kj}^{(k)} = -a_{kj}^{(k-1)}$ and $r_{ik}^{(k)} = -a_{ik}^{(k-1)}$. The lower triangular matrix $L^{(k)}$ is the identity, except for the kth column, which is

$$\left(0,\dots,0,1,-\frac{\tilde{a}_{k+1,k}^{(k)}}{\tilde{a}_{kk}^{(k)}},\dots,-\frac{\tilde{a}_{nk}^{(k)}}{\tilde{a}_{kk}^{(k)}}\right)^T.$$

From this it is easily seen that $A^{(k)}$ is the matrix that arises from $\tilde{A}^{(k)}$ by eliminating elements in the kth column using row k, while $\tilde{A}^{(k)}$ is obtained from $A^{(k-1)}$ by replacing entries in row or column k whose indices are in P by 0.

Now, $A^{(0)} = A$ is an M-matrix, so $R^{(1)} \geq 0$. From Lemma 11.1.2 it follows that $\tilde{A}^{(1)}$ is an M-matrix and, therefore, $L^{(1)} \geq 0$. From Lemma 11.1.1 it follows that $A^{(1)}$ is an M-matrix. Continuing the argument in this fashion, we can prove that $A^{(k)}$ and $\tilde{A}^{(k)}$ are M-matrices and $L^{(k)} \geq 0$ and $R^{(k)} \geq 0$ for $k = 1,\dots,n-1$. From the definitions it follows immediately that

$$
\begin{aligned}
L^{(k)}R^{(m)} &= R^{(m)} \text{ if } k < m, \\
A^{(n-1)} &= L^{(n-1)}\tilde{A}^{(n-1)} = L^{(n-1)}A^{(n-2)} + L^{(n-1)}R^{(n-1)} = \cdots \\
&= \left(\prod_{j=1}^{n-1} L^{(n-j)}\right) A^{(0)} + \sum_{i=1}^{n-1}\left(\prod_{j=1}^{n-i} L^{(n-j)}\right) R^{(i)}.
\end{aligned}
$$

By combining these equations we have

$$A^{(n-1)} = \left(\prod_{j=1}^{n-1} L^{(n-j)} \right) \left(A + \sum_{i=1}^{n-1} R^{(i)} \right).$$

Let us now define $U \equiv A^{(n-1)}$, $L \equiv (\prod_{j=1}^{n-1} L^{(n-j)})^{-1}$, and $R \equiv \sum_{i=1}^{n-1} R^{(i)}$. Then $LU = A + R$, $(LU)^{-1} \geq 0$, and $R \geq 0$, so the splitting $A = LU - R$ is regular. The uniqueness of the factors L and U follows from equating the elements of A and LU for $(i, j) \notin P$ and from the fact that L has a unit diagonal. $\square$

COROLLARY 11.1.1 (Meijerink and van der Vorst). *If A is a symmetric M-matrix, then for each subset P of the off-diagonal indices with the property that $(i, j) \in P$ implies $(j, i) \in P$, there exists a unique lower triangular matrix L with $l_{ij} = 0$ if $(i, j) \in P$ such that $A = LL^T - R$, where $r_{ij} = 0$ if $(i, j) \notin P$. The splitting $A = LL^T - R$ is a regular splitting.*

When A has the sparsity pattern of the 5-point approximation to the diffusion equation (9.6–9.8), the incomplete Cholesky decomposition that forces L to have the same sparsity pattern as the lower triangle of A is especially simple. It is convenient to write the incomplete decomposition in the form LDL^T, where D is a diagonal matrix. Let $\mathbf{a}$ denote the main diagonal of A, $\mathbf{b}$ the first lower diagonal, and $\mathbf{c}$ the $(m + 1)$st lower diagonal, where $m = n_x$. Let $\tilde{\mathbf{a}}$ denote the main diagonal of L, $\tilde{\mathbf{b}}$ the first lower diagonal, and $\tilde{\mathbf{c}}$ the $(m + 1)$st lower diagonal; let $\tilde{\mathbf{d}}$ denote the main diagonal of D. Then we have

$$\tilde{\mathbf{b}} = \mathbf{b}, \quad \tilde{\mathbf{c}} = \mathbf{c},$$

$$\tilde{\mathbf{a}}_i = \tilde{\mathbf{d}}_i^{-1} = \mathbf{a}_i - \tilde{\mathbf{b}}_{i-1}^2 \tilde{\mathbf{d}}_{i-1} - \tilde{\mathbf{c}}_{i-m}^2 \tilde{\mathbf{d}}_{i-m}, \quad i = 1 \cdots n.$$

The product $M = LDL^T$ has an ith row of the form

$$\cdots \quad 0 \quad \mathbf{c}_{i-m} \quad \mathbf{r}_{i-m+1} \quad 0 \quad \cdots \quad 0 \quad \mathbf{b}_{i-1} \quad \mathbf{a}_i \quad \mathbf{b}_i \quad 0 \quad \cdots \quad 0 \quad \mathbf{r}_i \quad \mathbf{c}_i \quad 0 \quad \cdots,$$

where $\mathbf{r}_i = (\mathbf{b}_{i-1} \mathbf{c}_{i-1})/\tilde{\mathbf{a}}_{i-1}$. Usually the off-diagonal entries $\mathbf{b}_{i-1}$ and $\mathbf{c}_{i-1}$ are significantly smaller in absolute value than $\mathbf{a}_i$ (for the model problem, $\mathbf{b}_{i-1} \mathbf{c}_{i-1}/\mathbf{a}_i = 1/4$) and are also significantly smaller in absolute value than $\tilde{\mathbf{a}}_i$. Thus, one expects the remainder matrix R in the splitting $A = M - R$ to be small in comparison to A or M.

Although the incomplete Cholesky decomposition is a regular splitting, it cannot be compared to preconditioners such as the diagonal of A or the lower triangle of A (using Corollary 10.3.1 or Theorem 10.4.1), because some entries of the incomplete Cholesky preconditioner $M = LDL^T$ are closer to those of A than are the corresponding entries of $\text{diag}(A)$ or lower triangle(A), but some entries are further away. Numerical evidence suggests, however, that the incomplete Cholesky preconditioner used with the CG algorithm often requires significantly fewer iterations than a simple diagonal preconditioner. Of course, each iteration requires somewhat more work, and backsolving with the incomplete Cholesky factors is not an easily parallelizable operation. Consequently, there have been a number of experiments suggesting that on vector or parallel computers it may be faster just to use $M = \text{diag}(A)$ as a preconditioner.

Other sparsity patterns can be used for the incomplete Cholesky factors. For example, while the previously described preconditioner is often referred to as IC(0) since the factor L has no diagonals that are not already in A, Meijerink and van der Vorst suggest the preconditioner IC(3), where the set P of zero off-diagonal indices is

$$P = \{(i,j) \ : \ |i - j| \neq 0, \ 1, \ 2, \ m - 2, \ m - 1, \ m\}.$$

With this preconditioner, L has three extra nonzero diagonals—the second, $(m-2)$nd, and $(m-1)$st subdiagonals—and again the entries are chosen so that LDL^T matches A in positions not in P.

The effectiveness of the incomplete Cholesky decomposition as a preconditioner depends on the *ordering* of equations and unknowns. For example, with the red–black ordering of nodes for the model problem, the matrix A takes the form (9.9), where D_1 and D_2 are diagonal and B, which represents the coupling between red and black points, is also sparse. With this ordering, backsolving with the IC(0) factor is more parallelizable than for the natural ordering since L takes the form

$$L = \begin{pmatrix} \tilde{D}_1 & 0 \\ \tilde{B}^T & \tilde{D}_2 \end{pmatrix}.$$

One can solve a linear system with coefficient matrix L by first determining the red components of the solution in parallel, then applying the matrix $\tilde{B}^T$ to these components, and then solving for the black components in parallel. Unfortunately, however, the incomplete Cholesky preconditioner obtained with this ordering is significantly *less* effective in reducing the number of CG iterations required than that obtained with the natural ordering.

11.2. Modified Incomplete Cholesky Decomposition.

While the incomplete Cholesky preconditioner may significantly reduce the number of iterations required by the PCG algorithm, we will see in this section that for second-order elliptic differential equations the number of iterations is still $O(h^{-1})$, as it is for the unpreconditioned CG algorithm; that is, the condition number of the preconditioned matrix is $O(h^{-2})$. Only the constant has been improved. A slight modification of the incomplete Cholesky decomposition, however, can lead to an $O(h^{-1})$ condition number. Such a modification was developed by Dupont, Kendall, and Rachford [37] and later by Gustafsson [74]. Also, see [6]. The main results of this section are from [74].

Consider the matrix $A \equiv A_h$ in (9.6–9.8) arising from the 5-point approximation to the steady-state diffusion equation, or, more generally, consider any matrix A_h obtained from a finite difference or finite element approximation with mesh size h for the second-order self-adjoint elliptic differential equation

$$(11.1) \qquad \mathcal{L}u \equiv -\frac{\partial}{\partial x}\left(\alpha_1 \frac{\partial u}{\partial x}\right) - \frac{\partial}{\partial y}\left(\alpha_2 \frac{\partial u}{\partial y}\right) = f$$

defined on a region $\Omega \subset \mathbf{R}^2$, with appropriate boundary conditions on $\partial\Omega$. Assume $\alpha_i \equiv \alpha_i(x, y) \geq \alpha > 0$, $i = 1, 2$.

Such a matrix usually has several special properties. First, it contains only *local couplings* in the sense that if $a_{ij} \neq 0$, then the distance from node i to node j is bounded by a constant (independent of h) times h. We will write this as $O(h)$. Second, since each element of a matrix vector product Av approximates $\mathcal{L}v(x, y)$, where $v(x, y)$ is the function represented by the vector v, and since $\mathcal{L}$ acting on a constant function v yields 0, the *row sums* of A are zero, except possibly at points that couple to the boundary of Ω. Assume that A is scaled so that the nonzero entries of A are of size $O(1)$. The dimension n of A is $O(h^{-2})$. The 5-point Laplacian (multiplied by h^2) is a typical example:

$$(11.2) \qquad A = \begin{pmatrix} T & -I & & \\ -I & T & \ddots & \\ & \ddots & \ddots & -I \\ & & -I & T \end{pmatrix}, \qquad T = \begin{pmatrix} 4 & -1 & & \\ -1 & 4 & \ddots & \\ & \ddots & \ddots \end{pmatrix}.$$

If $A = M - R$ is a splitting of A, then the largest and smallest eigenvalues of the preconditioned matrix $M^{-1}A$ are

$$\max_{v \neq 0} \frac{\langle Av, v \rangle}{\langle Mv, v \rangle} \quad \text{and} \quad \min_{v \neq 0} \frac{\langle Av, v \rangle}{\langle Mv, v \rangle},$$

and $\langle Av, v \rangle / \langle Mv, v \rangle$ can be written in the form

$$(11.3) \qquad \frac{\langle Av, v \rangle}{\langle Mv, v \rangle} = \frac{1}{1 + \langle Rv, v \rangle / \langle Av, v \rangle}.$$

Suppose the vector v represents a function $v(x, y)$ in $C_0^1(\Omega)$—the space of continuously differentiable functions with value 0 on the boundary of Ω. By an elementary summation by parts, we can write

$$(11.4) \qquad \langle Av, v \rangle = -\sum_i \sum_{j>i} a_{ij}(v_i - v_j)^2 + \sum_i \sum_j a_{ij} v_i^2.$$

Because of the zero row sum property of A, we have $\sum_j a_{ij} v_i^2 = 0$ unless node i is coupled to the boundary of Ω, and this happens only if the distance from node i to the boundary is $O(h)$. Since $v(x, y) \in C_0^1(\Omega)$, it follows that at such points $|v_i|$ is bounded by $O(h)$. Consequently, since the nonzero entries of A are of order $O(1)$, the second sum in (11.4) is bounded in magnitude by the number of nodes i that couple to $\partial\Omega$ times $O(h^2)$. In most cases this will be $O(h)$.

Because of the local property of A, it follows that for nodes i and j such that a_{ij} is nonzero, the distance between nodes i and j is $O(h)$ and, therefore, $|v_i - v_j|$ is bounded by $O(h)$. The first sum in (11.4) therefore satisfies

$$\left|\sum_i \sum_{j>i} a_{ij}(v_i - v_j)^2\right| \le O(1)$$

since there are $O(h^{-2})$ terms, each of size $O(h^2)$.

For the remainder matrix R, we can also write

(11.5) $$\langle Rv, v\rangle = -\sum_i \sum_{j>i} r_{ij}(v_i - v_j)^2 + \sum_i \sum_j r_{ij}v_i^2.$$

Suppose that the remainder matrix also has the property that nonzero entries r_{ij} correspond only to nodes i and j that are separated by no more than $O(h)$, and suppose also that the nonzero entries of R are of size $O(1)$ (but are perhaps smaller than the nonzero entries of A). This is the case for the incomplete Cholesky decomposition where, for the 5-point Laplacian, r_{ij} is nonzero only if $j = i + m - 1$ or $j = i - m + 1$. These positions correspond to the nodes pictured below, whose distance from node i is $\sqrt{2}h$.

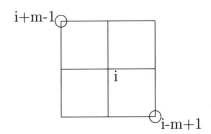

Then, by the same argument as used for A, the first sum in (11.5) is bounded in absolute value by $O(1)$.

The bound on the second term in (11.4), however, depended on the zero row sum property of A. If this property is not shared by R (and it is not for the incomplete Cholesky decomposition or for *any* regular splitting, since the entries of R are all nonnegative), then this second sum could be much larger. It is bounded by the number of nonzero entries of R in rows corresponding to nodes away from the boundary, which is typically $O(h^{-2})$, times the nonzero values of r_{ij}, which are of size $O(1)$, times the value of the function $v(x,y)$ away from the boundary, which is $O(1)$. Hence the second sum in (11.5) may be as large as $O(h^{-2})$. For vectors v representing a $\mathbf{C}_0^1$-function, the ratio $\langle Rv, v\rangle/\langle Av, v\rangle$ in (11.3) is then of size $O(h^{-2})$, so if $\langle Rv, v\rangle$ is positive (as it is for a regular splitting if $v \ge 0$), then the ratio $\langle Av, v\rangle/\langle Mv, v\rangle$ in (11.3) is of size $O(h^2)$. In contrast, if we consider the first unit vector ξ_1, for example, then $\langle A\xi_1, \xi_1 \rangle / \langle M\xi_1, \xi_1 \rangle = O(1)$. It follows that the condition number of the preconditioned matrix is at least $O(h^{-2})$, which is the same order as $\kappa(A)$.

We therefore seek a preconditioner $M = LL^T$ such that $A = M - R$ and $|\langle Rv, v \rangle| \leq O(h^{-1})$ for $v(x, y) \in \mathbf{C}_0^1$, in order to have a chance of producing a preconditioned matrix with condition number $O(h^{-1})$ instead of $O(h^{-2})$. Suppose A is written in the form $A = M - R$, where

$$(11.6) \qquad\qquad R = \hat{R} + E$$

and where $\hat{R}$ is negative semidefinite (that is, $\langle \hat{R}v, v \rangle \leq 0 \ \forall v$), $\sum_j \hat{r}_{ij} = 0 \ \forall i$, and E is a positive definite diagonal matrix. Assume also that $\hat{R}$ has nonzero entries only in positions (i, j) corresponding to nodes i and j that are within $O(h)$ of each other. Our choice of the matrix E depends on the boundary conditions. For Dirichlet problems, which will be dealt with here, we choose $E = \eta h^2 \mathrm{diag}(A)$, where $\eta > 0$ is a parameter. For Neumann and mixed problems, similar results can be proved if some elements of E, corresponding to points on the part of the boundary with Neumann conditions, are taken to be of order $O(h)$.

From (11.5), it can be seen that R in (11.6) satisfies

$$\langle Rv, v \rangle = \sum_i \sum_j r_{ij} |v_i|^2 + O(1)$$

when $v(x, y) \in \mathbf{C}_0^1(\Omega)$, since the first sum in (11.5) is of size $O(1)$. Since the row sums of $\hat{R}$ are all zero and the nonzero entries of E are of size $O(h^2)$, we have

$$\langle Rv, v \rangle = O(1),$$

so the necessary condition $|\langle Rv, v \rangle| \leq O(h^{-1})$ is certainly satisfied. The following theorem gives a *sufficient* condition to obtain a preconditioned matrix with condition number $O(h^{-1})$.

THEOREM 11.2.1 (Gustafsson). *Let $A = M - R$, where R is of the form (11.6), $\hat{R}$ is negative semidefinite and has zero row sums and only local couplings, and E is a positive definite diagonal matrix with diagonal entries of size $O(h^2)$. Then a sufficient condition to obtain $\lambda_{max}(M^{-1}A)/\lambda_{min}(M^{-1}A) = O(h^{-1})$ is*

$$(11.7) \qquad\qquad 0 \leq -\langle \hat{R}v, v \rangle \leq (1 + ch)^{-1} \langle Av, v \rangle \ \forall v,$$

where $c > 0$ is independent of h.

Proof. There exist constants c_1 and c_2, independent of h, such that $c_1 h^2 \leq \langle Av, v \rangle / \langle v, v \rangle \leq c_2$. Since the entries of E are of order h^2, it follows that $0 < \langle Ev, v \rangle / \langle Av, v \rangle \leq c_3$ for some constant c_3. From (11.3) and the fact that E is positive definite and $\hat{R}$ is negative semidefinite, we can write

$$(1 + c_3)^{-1} \leq \frac{1}{1 + \langle Ev, v \rangle / \langle Av, v \rangle} \leq \frac{\langle Av, v \rangle}{\langle Mv, v \rangle} \leq \frac{1}{1 + \langle \hat{R}v, v \rangle / \langle Av, v \rangle}.$$

The rightmost expression here, and hence $\lambda_{max}(M^{-1}A)/\lambda_{min}(M^{-1}A)$, is of order $O(h^{-1})$ if $\hat{R}$ satisfies (11.7). $\quad\square$

When A is an M-matrix arising from discretization of (11.1), a simple modification of the incomplete Cholesky idea, known as *modified incomplete Cholesky decomposition* (MIC) [37, 74], yields a preconditioner M such that $\lambda_{max}(M^{-1}A)/\lambda_{min}(M^{-1}A) = O(h^{-1})$. Let L be a lower triangular matrix with zeros in positions corresponding to indices in some set P. Choose the nonzero entries of L so that $M = LL^T$ matches A in positions outside of P except for the main diagonal. Setting $E = \eta h^2 \text{diag}(A)$, also force $\hat{R} \equiv LL^T - (A + E)$ to have zero rowsums. It can be shown, similar to the unmodified incomplete Cholesky case, that this decomposition exists for a general M-matrix A and that the off-diagonal elements of $\hat{R}$ are nonnegative while the diagonal elements are negative. As for ordinary incomplete Cholesky decomposition, a popular choice for the set P is the set of positions in which A has zeros, so L has the same sparsity pattern as the lower triangle of A.

When A has the sparsity pattern of the 5-point approximation (9.6–9.8), this can be accomplished as follows. Again, it is convenient to write the modified incomplete Cholesky decomposition in the form LDL^T, where D is a diagonal matrix. Let $\mathbf{a}$ denote the main diagonal of A, $\mathbf{b}$ the first lower diagonal, and $\mathbf{c}$ the $(m + 1)$st lower diagonal. Let $\tilde{\mathbf{a}}$ denote the main diagonal of L, $\tilde{\mathbf{b}}$ the first lower diagonal, and $\tilde{\mathbf{c}}$ the $(m+1)$st lower diagonal; let $\tilde{\mathbf{d}}$ denote the main diagonal of D. Then we have $\tilde{\mathbf{b}} = \mathbf{b}$, $\tilde{\mathbf{c}} = \mathbf{c}$, and for $i = 1, \ldots, n$,

$$(11.8)\,\tilde{\mathbf{a}}_i = \tilde{\mathbf{d}}_i^{-1} = \mathbf{a}_i(1 + \eta h^2) - \tilde{\mathbf{b}}_{i-1}^2 \tilde{\mathbf{d}}_{i-1} - \tilde{\mathbf{c}}_{i-m}^2 \tilde{\mathbf{d}}_{i-m} - \mathbf{r}_{i-1} - \mathbf{r}_{i-m},$$

$$(11.9)\,\mathbf{r}_i = \tilde{\mathbf{b}}_i \tilde{\mathbf{c}}_i \tilde{\mathbf{d}}_i,$$

where elements not defined should be replaced by zeros. The matrix $\hat{R}$ in (11.6) satisfies

$$\hat{r}_{i+1,i+m} = \hat{r}_{i+m,i+1} = \mathbf{r}_i, \quad \hat{r}_{i+1,i+1} = -\mathbf{r}_i - \mathbf{r}_{i+1-m},$$

and all other elements of $\hat{R}$ are zero.

It can be shown that for smooth coefficients $\alpha_1(x, y)$ and $\alpha_2(x, y)$, the above procedure yields a preconditioner M for which the preconditioned matrix $L^{-1}AL^{-T}$ has condition number $O(h^{-1})$. For simplicity, we will show this only for the case when $\alpha_1(x, y) \equiv \alpha_2(x, y) \equiv 1$ and A is the standard 5-point Laplacian (11.2). The technique of proof is similar in the more general case.

LEMMA 11.2.1 (Gustafsson). *Let* $\mathbf{r}_i$, $i = 1, \ldots, n - m$, *be the elements defined by* (11.8–11.9) *for the 5-point Laplacian matrix* A. *Then*

$$0 \leq \mathbf{r}_i \leq \frac{1}{2(1 + ch)},$$

where $c > 0$ *is independent of* h.

Proof. We first show that

$$\tilde{\mathbf{a}}_i \geq 2(1 + \sqrt{2\eta}\,h) \quad \forall i.$$

For the model problem, the recurrence equations (11.8–11.9) can be written in the form

$$\tilde{\mathbf{a}}_i = 4(1 + \eta h^2) - \mathbf{b}_{i-1}(\mathbf{b}_{i-1} + \mathbf{c}_{i-1})/\tilde{\mathbf{a}}_{i-1} - \mathbf{c}_{i-m}(\mathbf{c}_{i-m} + \mathbf{b}_{i-m})/\tilde{\mathbf{a}}_{i-m}.$$

For $i = 1$, we have $\tilde{\mathbf{a}}_1 = 4(1 + \eta h^2) \geq 2(1 + \sqrt{2\eta}\, h)$. Assume that $\tilde{\mathbf{a}}_j \geq 2(1 + \sqrt{2\eta}\, h)$ for $j = 1, \ldots, i - 1$. Then we have

$$\begin{aligned}
\tilde{\mathbf{a}}_i &\geq 4(1 + \eta h^2) - 2/(1 + \sqrt{2\eta}\, h) \\
&\geq 4(1 + \eta h^2) - 2 \cdot (1 - \sqrt{2\eta}\, h + 2\eta h^2) \\
&\geq 2 + 2\sqrt{2\eta}\, h.
\end{aligned}$$

(In fact, for n sufficiently large, the elements $\tilde{\mathbf{a}}_i$ approach a constant value γ satisfying

$$\gamma = 4(1 + \eta h^2) - 4/\gamma.$$

The value is $\gamma = 2(1 + \sqrt{2\eta + \eta^2 h} + \eta h^2).$)

Since $\mathbf{r}_i = \mathbf{b}_i \mathbf{c}_i / \tilde{\mathbf{a}}_i$, we obtain

$$0 \leq \mathbf{r}_i \leq \frac{1}{2(1 + \sqrt{2\eta}\, h)}. \qquad \square$$

THEOREM 11.2.2 (Gustafsson). *Let* $M = LL^T$, *where the nonzero elements of* L *are defined by* (11.8–11.9), *and* A *is the 5-point Laplacian matrix. Then* $\lambda_{max}(M^{-1}A)/\lambda_{min}(M^{-1}A) = O(h^{-1})$.

Proof. For the model problem, using expression (11.4), we can write

$$(11.10) \qquad \langle Av, v \rangle \geq - \sum_i [\mathbf{b}_i(v_i - v_{i+1})^2 + \mathbf{c}_i(v_i - v_{i+m})^2]$$

for any vector v. An analogous expression for $\langle \hat{R}v, v \rangle$ shows, since the row sums of $\hat{R}$ are all zero,

$$\langle \hat{R}v, v \rangle = - \sum_i \sum_{j > i} \hat{r}_{ij}(v_i - v_j)^2 = - \sum_i \mathbf{r}_{i-1}(v_i - v_{i-1+m})^2.$$

Since $-\hat{R}$ is a symmetric weakly diagonally dominant matrix with nonnegative diagonal elements and nonpositive off-diagonal elements, it follows that $\hat{R}$ is negative semidefinite. From Lemma 11.2.1 it follows that

$$(11.11) \qquad -\langle \hat{R}v, v \rangle \leq \frac{1}{2(1 + ch)} \sum_{i : \mathbf{r}_{i-1} \neq 0} (v_i - v_{i-1+m})^2.$$

Using the inequality $\frac{1}{2}(a - b)^2 \leq (a - c)^2 + (c - b)^2$, which holds for any real numbers a, b, and c, inequality (11.11) can be written in the form

$$-\langle \hat{R}v, v \rangle \leq \frac{1}{1 + ch} \sum_{i : \mathbf{r}_{i-1} \neq 0} [(v_i - v_{i-1})^2 + (v_{i-1} - v_{i-1+m})^2],$$

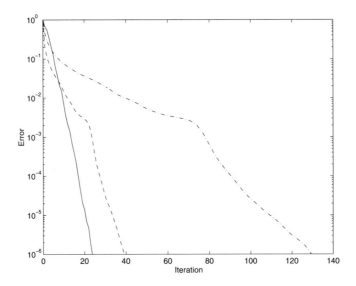

FIG. 11.1. *Convergence of iterative methods for the model problem, $h = 1/51$. Unpreconditioned CG (dash–dot), ICCG (dashed), MICCG (solid).*

where the right-hand side can also be expressed as $(1 + ch)^{-1} \sum_{i:\mathbf{r}_i \neq 0} [(v_{i+1} - v_i)^2 + (v_i - v_{i+m})^2]$. Since $\mathbf{r}_i$ is nonzero only when $\mathbf{b}_i$ and $\mathbf{c}_i$ are nonzero, we combine this with inequality (11.10) and obtain

$$-\langle \hat{R}v, v \rangle \leq (1 + ch)^{-1} \langle Av, v \rangle.$$

The desired result then follows from Lemma 11.2.1.   □

   For sufficiently small values of h, it is clear that MIC(0) gives a better condition number for the preconditioned system than does IC(0). In fact, even for coarse grids, the MIC(0) preconditioner, with a small parameter η, gives a significantly better condition number than IC(0) for the model problem. Figure 11.1 shows the convergence of unpreconditioned CG, ICCG(0), and MICCG(0) for the model problem with $h = 1/51$. The quantity plotted is the 2-norm of the error divided by the 2-norm of the true solution, which was set to a random vector. A zero initial guess was used. The parameter η in the *MIC* preconditioner was set to .01, although the convergence behavior is not very sensitive to this parameter. For this problem the condition numbers of the iteration matrices are as follows: unpreconditioned, 1053; IC(0), 94; MIC(0), 15. Although the bound (3.8) on the A-norm of the error in terms of the square root of the condition number may be an overestimate of the actual A-norm of the error, one does find that as the mesh is refined, the number of unpreconditioned CG and ICCG iterations tends to grow like $O(h^{-1})$, while the number of MICCG iterations grows like $O(h^{-1/2})$.

   When ICCG and MICCG are applied in practice to problems other than the model problem, it has sometimes been observed that ICCG actually converges faster, despite a significantly larger condition number. This might be accounted for by a smaller sharp error bound (3.6) for ICCG, but the reason appears to be that rounding errors have a greater effect on the convergence rate of MICCG, because of more large, well-separated eigenvalues. For a discussion, see [133].

Comments and Additional References.

Sometimes the set P of zero entries in the (modified) incomplete Cholesky or incomplete LU decomposition is not set ahead of time, but, instead, entries are discarded only if their absolute values lie below some threshold. See, for example, [100]. Further analysis of incomplete factorizations can be found in a number of places, including [6, 7, 8, 13, 14, 106].

In addition to incomplete LU decompositions, incomplete QR factorizations have been developed and used as preconditioners [116]. In order to make better use of parallelism, sparse approximate inverses have also been proposed as preconditioners. See, for instance, [15, 16, 17, 88].

The analysis given here for modified incomplete Cholesky decomposition applied to the model problem and the earlier analysis of the SOR method for the model problem were not so easy. The two methods required very different proof techniques, and similar analysis for other preconditioners would require still different arguments. If one changes the model problem slightly, however, by replacing the Dirichlet boundary conditions by *periodic* boundary conditions, then the analysis of these and other preconditioners becomes *much* easier. The reason is that the resulting coefficient matrix and preconditioners all have the same Fourier modes as eigenvectors. Knowing the eigenvalues of the coefficient matrix and the preconditioner, it then becomes relatively easy to identify the largest and smallest ratios, which are the extreme eigenvalues of the preconditioned matrix. It has been observed numerically and argued heuristically that the results obtained for the periodic problem are very similar to those for the model problem with Dirichlet boundary conditions. For an excellent discussion, see [24].

Chapter 12

Multigrid and Domain Decomposition Methods

Chapters 10 and 11 dealt with preconditioners designed for general classes of matrices. The origin of the problem was not a factor in defining the preconditioner, although analysis of the preconditioner was sometimes limited to problems arising from certain types of partial differential equations. In this chapter we deal with preconditioners designed specifically for problems arising from partial differential equations. The methods are intended for use with broad classes of problems and are not restricted to one particular equation. Attempts have been made to extend some of these ideas to general linear systems, as in algebraic multigrid methods, but the extensions are not immediate.

Because the methods are designed for partial differential equation problems, their analysis may require detailed knowledge of the properties of finite difference and finite element approximations, while we are assuming just a basic familiarity with these ideas. For this reason, we restrict our analysis to the model problem, where properties shared by more general finite element and finite difference approximations can be verified directly, and we indicate how the analysis can be extended.

12.1. Multigrid Methods.

Multigrid methods were not originally described as a combination of an iteration scheme and a preconditioner, but it is clear that they can be viewed in this way. The first multigrid methods used simple iteration, so we will start with that approach. It will be apparent that the same preconditioners can be used with any of the other Krylov subspace methods described in Part I of this book.

Recall the simple iteration scheme

$$(12.1) \qquad x_k = x_{k-1} + M^{-1}(b - Ax_{k-1}).$$

The error $e_k \equiv A^{-1}b - x_k$ is given by

$$(12.2) \qquad e_k = (I - M^{-1}A)e_{k-1},$$

so the norm of the error satisfies

$$\|e_k\| \le \|I - M^{-1}A\| \cdot \|e_{k-1}\|.$$

The error is reduced quickly if $\|I - M^{-1}A\| << 1$.

Most multigrid methods can be written in the general form (12.1), where the iterates x_k represent quantities generated after a *coarse grid correction cycle* and a given number of *relaxation sweeps*. That is, given an approximation x_{k-1} to the solution, the multigrid algorithm generates a new approximation $x_{k-1,0}$ via a formula of the form

$$(12.3) \qquad x_{k-1,0} = x_{k-1} + C(b - Ax_{k-1}),$$

where the matrix C represents a *coarse grid* approximation to A^{-1}. The method then generates a certain number, say ℓ, of new approximations, $x_{k-1,j}$, $j = 1, \ldots, \ell$ by performing *relaxation sweeps* of the form

$$(12.4) \qquad x_{k-1,j} = x_{k-1,j-1} + G(b - Ax_{k-1,j-1}), \quad j = 1, \ldots, \ell,$$

where the matrix G also represents an approximation to A^{-1}. If we denote by x_k the quantity $x_{k-1,\ell}$, then we find

$$x_k = x_{k-1} + [I - (I - GA)^\ell(I - CA)A^{-1}](b - Ax_{k-1}),$$

and the error e_k satisfies

$$(12.5) \qquad e_k = (I - GA)^\ell(I - CA)e_{k-1}.$$

Thus, for multigrid methods, the matrix $I - M^{-1}A$ in (12.2) is of the special form $(I - GA)^\ell(I - CA)$ for certain matrices C and G.

12.1.1. Aggregation Methods.

We first analyze iterations of the form (12.3–12.4) in a general setting, where the matrix C involves the inverse of a smaller matrix. Such methods are sometimes called *aggregation methods*.

While multigrid and aggregation methods can be applied to non-Hermitian and indefinite problems, the analysis here will be restricted to Hermitian positive definite problems. We will estimate the rate at which the A-norm of the error, $\|e_k\|_A \equiv \langle e_k, Ae_k \rangle^{1/2}$, is reduced. Taking norms on each side in equation (12.5), we find that

$$\|e_k\|_A \le \|(I - GA)^\ell(I - CA)\|_A \cdot \|e_{k-1}\|_A.$$

The quantity $\|(I - GA)^\ell(I - CA)\|_A$ is called the *contraction number* of the method and will be denoted by σ. In terms of the 2-norm, σ is given by

$$(12.6) \qquad \sigma = \|(I - A^{1/2}GA^{1/2})^\ell(I - A^{1/2}CA^{1/2})\|.$$

A simple bound for σ is

$$(12.7) \qquad \sigma \le \|I - A^{1/2}GA^{1/2}\|^\ell \cdot \|I - A^{1/2}CA^{1/2}\|.$$

The methods to be considered use matrices C and G, for which $\|I - A^{1/2}CA^{1/2}\| = 1$ and $\|I - A^{1/2}GA^{1/2}\| < 1$. Hence the methods are *convergent* whenever A is Hermitian and positive definite, and, moreover, they reduce the A-norm of the error at each step.

Inequality (12.7), however, is too crude an estimate to provide much useful information about the rate of convergence. In fact, the methods to be considered use matrices C and G, which are designed to complement each other in such a way that the norm of the matrix product in (12.6) is much less than the product of the norms in (12.7). Instead of inequality (12.7), we use the definition of the matrix norm to estimate σ by

$$(12.8) \qquad \sigma \leq \max_{\substack{\|y\| = \|I - A^{1/2}CA^{1/2}\| \\ y \in \text{range}(I - A^{1/2}CA^{1/2})}} \|(I - A^{1/2}GA^{1/2})^{\ell}y\|.$$

If the range of $I - A^{1/2}CA^{1/2}$ is a restricted set of vectors on which $I - A^{1/2}GA^{1/2}$ is highly contractive, then the bound in (12.8) may be much smaller than that in (12.7).

We now define the form of the matrix C in iteration (12.3). Suppose A is an n-by-n matrix and $\hat{n} < n$. Let $I_{\hat{n}}^n$ be an arbitrary n-by-$\hat{n}$ matrix of rank $\hat{n}$, and define an $\hat{n}$-by-n matrix $I_n^{\hat{n}}$ by

$$(12.9) \qquad\qquad I_n^{\hat{n}} = (I_{\hat{n}}^n)^H.$$

Define an $\hat{n}$-by-$\hat{n}$ matrix $\hat{A}$ by

$$(12.10) \qquad\qquad \hat{A} = I_n^{\hat{n}} A I_{\hat{n}}^n,$$

and take C to be the matrix

$$(12.11) \qquad\qquad C = I_{\hat{n}}^n \hat{A}^{-1} I_n^{\hat{n}}.$$

The following theorem shows that when C is defined in this way, the matrix $A^{1/2}CA^{1/2}$ is just the orthogonal projector from $\mathbf{C}^n$ to the $\hat{n}$-dimensional subspace $A^{1/2} \cdot \text{range}(I_{\hat{n}}^n)$.

THEOREM 12.1.1. *If C is defined by (12.9–12.11), then*

$$(12.12) \qquad\qquad A^{1/2} \cdot \mathcal{R}(I - A^{1/2}CA^{1/2}) \subseteq \mathcal{N}(I_n^{\hat{n}}),$$

where $\mathcal{R}(\cdot)$ denotes the range and $\mathcal{N}(\cdot)$ the null space of an operator. The matrix $A^{1/2}CA^{1/2}$ is an orthogonal projector from $\mathbf{C}^n$ to an $\hat{n}$-dimensional subspace and hence

$$(12.13) \qquad\qquad \|I - A^{1/2}CA^{1/2}\| = 1.$$

Proof. Using definitions (12.10) and (12.11), we find

$$I_n^{\hat{n}} A^{1/2}(I - A^{1/2}CA^{1/2}) = I_n^{\hat{n}} A^{1/2} - (I_n^{\hat{n}} A I_{\hat{n}}^n)A_{\hat{n}}^{-1} I_n^{\hat{n}} A^{1/2} = 0.$$

This establishes (12.12). To establish (12.13), note that since $I - A^{1/2}CA^{1/2}$ is a Hermitian matrix, its norm is the absolute value of its largest eigenvalue. If z is an eigenvector of this matrix with eigenvalue λ, then, using (12.12), we can write

$$I_n^{\hat{n}} A^{1/2}(I - A^{1/2}CA^{1/2})z = 0 = \lambda I_n^{\hat{n}} A^{1/2} z.$$

It follows that either $\lambda = 0$ or $A^{1/2}z \in \mathcal{N}(I_n^{\hat{n}})$. In the latter case, $CA^{1/2}z$ is zero and hence $\lambda = 1$. Since the eigenvalues of $I - A^{1/2}CA^{1/2}$, and hence of $A^{1/2}CA^{1/2}$, are 0's and 1's, this establishes that $A^{1/2}CA^{1/2}$ is an orthogonal projector and that (12.13) holds.   □

Applying the theorem, inequality (12.8) becomes

$$\sigma \leq \max_{\substack{\|y\| = 1 \\ A^{1/2}y \in \mathcal{N}(I_n^{\hat{n}})}} \|(I - A^{1/2}GA^{1/2})^\ell y\|,$$

and since the null space of $I_n^{\hat{n}}$ is the orthogonal complement of the range of $I_{\hat{n}}^n$, this can be written as

(12.14)
$$\sigma \leq \max_{\substack{\|y\| = 1 \\ y \perp A^{1/2} \cdot \mathcal{R}(I_{\hat{n}}^n)}} \|(I - A^{1/2}GA^{1/2})^\ell y\|.$$

Suppose the matrix G is given. Let $d_1^2 \geq \cdots \geq d_n^2$ denote the eigenvalues of $(I - A^{1/2}GA^{1/2})^{\ell H}(I - A^{1/2}GA^{1/2})^\ell$, and let $v_1, \ldots, v_n$ denote the corresponding orthonormal eigenvectors. For any vector y we can write $y = \sum_{i=1}^n \langle y, v_i \rangle v_i$ and

(12.15)
$$\|(I - A^{1/2}GA^{1/2})^\ell y\|^2 = \sum_{i=1}^n \langle y, v_i \rangle^2 d_i^2.$$

Now, in general, we have

(12.16)
$$\max_{\|y\|=1} \|(I - A^{1/2}GA^{1/2})^\ell y\| = |d_1|,$$

but with the additional constraint $y \perp A^{1/2} \cdot \mathcal{R}(I_{\hat{n}}^n)$, a smaller bound may be attained. If $I_{\hat{n}}^n$ can be chosen so that $v_1, \ldots, v_{\hat{n}}$—the eigenvectors corresponding to the $\hat{n}$ largest eigenvalues—lie in the space $A^{1/2} \cdot \mathcal{R}(I_{\hat{n}}^n)$, then y will have no components in the direction of these eigenvectors and expression (12.15) can be replaced by

$$\|(I - A^{1/2}GA^{1/2})^\ell y\|^2 = \sum_{i=\hat{n}+1}^n \langle y, v_i \rangle^2 d_i^2.$$

Under these *ideal* conditions—$v_1, \ldots, v_{\hat{n}} \in A^{1/2} \cdot \mathcal{R}(I_{\hat{n}}^n)$—the bound (12.16) is replaced by

(12.17)
$$\max_{\substack{\|y\| = 1 \\ y \perp A^{1/2} \cdot \mathcal{R}(I_{\hat{n}}^n)}} \|(I - A^{1/2}GA^{1/2})^\ell y\| = |d_{\hat{n}+1}|.$$

As an example, suppose G is taken to be of the form

$$(12.18) \qquad\qquad G = \gamma I,$$

where the constant γ is chosen in an optimal or near optimal way. Then the eigenvectors $v_1, \ldots, v_n$ of $(I - A^{1/2}GA^{1/2})^{\ell^H}(I - A^{1/2}GA^{1/2})^\ell$ are just the eigenvectors of A, and the eigenvalues $d_1^2, \ldots, d_n^2$ of this matrix are each of the form $(1 - \gamma\lambda_i)^{2\ell}$ for some i, where $\lambda_1 \leq \cdots \leq \lambda_n$ are the eigenvalues of A. In this case, the bound (12.16) becomes

$$\max_{\|y\|=1} \|(I - \gamma A)^\ell y\| = \max\{|1 - \gamma\lambda_1|^\ell, |1 - \gamma\lambda_n|^\ell\}.$$

To minimize this bound, take $\gamma = 2/(\lambda_n + \lambda_1)$ and then

$$\max_{\|y\|=1} \|(I - \gamma A)^\ell y\| = \left(\frac{\lambda_n - \lambda_1}{\lambda_n + \lambda_1}\right)^\ell = \left(\frac{\kappa - 1}{\kappa + 1}\right)^\ell, \qquad \kappa = \frac{\lambda_n}{\lambda_1}.$$

This is the usual bound on the convergence rate for the method of steepest descent.

On the other hand, suppose some of the eigenvectors of A, say, those corresponding to the $\hat{n}$ smallest eigenvalues of A, lie in the desired space $A^{1/2} \cdot \mathcal{R}(I_{\hat{n}}^n)$. Then an improved bound like (12.17) holds, and this bound becomes

$$(12.19) \qquad \max_{\substack{\|y\| = 1 \\ y \perp A^{1/2} \cdot \mathcal{R}(I_{\hat{n}}^n)}} \|(I - \gamma A)^\ell y\| = \max\{|1 - \gamma\lambda_{\hat{n}+1}|^\ell, |1 - \gamma\lambda_n|^\ell\}.$$

To minimize this bound, take $\gamma = 2/(\lambda_n + \lambda_{\hat{n}+1})$, and (12.19) becomes

$$(12.20) \qquad \max_{\substack{\|y\| = 1 \\ y \perp A^{1/2} \cdot \mathcal{R}(I_{\hat{n}}^n)}} \|(I - \gamma A)^\ell y\| = \left(\frac{\lambda_n - \lambda_{\hat{n}+1}}{\lambda_n + \lambda_{\hat{n}+1}}\right)^\ell = \left(\frac{\hat{\kappa} - 1}{\hat{\kappa} + 1}\right)^\ell,$$

$$\hat{\kappa} = \frac{\lambda_n}{\lambda_{\hat{n}+1}}.$$

Thus, the effective condition number of A is reduced from $\kappa = \lambda_n/\lambda_1$ to $\hat{\kappa} = \lambda_n/\lambda_{\hat{n}+1}$. If the latter ratio is much smaller, as is typically the case when the matrix A approximates a self-adjoint elliptic differential operator, then much faster convergence is achieved by using a partial step of the form (12.3) than by iterating only with steps of the form (12.4).

12.1.2. Analysis of a Two-Grid Method for the Model Problem.
Recall that for the model problem $-\triangle u = f$ in the unit square with Dirichlet boundary conditions, the matrix A arising from a 5-point finite difference approximation on a grid of m-by-m interior points with spacing $h = 1/(m+1)$ has eigenvalues

$$(12.21) \quad \lambda_{i,j} = 4h^{-2}\left[\sin^2\left(\frac{i\pi h}{2}\right) + \sin^2\left(\frac{j\pi h}{2}\right)\right], \quad i, j = 1, \ldots, m,$$

and the (p,q)-components of the corresponding eigenvectors are

(12.22) $v^{(i,j)}_{p,q} = 2h \, \sin(ihp\pi) \, \sin(jhq\pi), \quad p,q = 1,\ldots,m,$

as shown in Theorem 9.1.2. The eigenvalues are all positive and the smallest and largest eigenvalues are given in Corollary 9.1.2:

(12.23) $\lambda_{min} = 8h^{-2} \sin^2 \left(\dfrac{\pi h}{2} \right) = 2\pi^2 + O(h^2),$

(12.24) $\lambda_{max} = 8h^{-2} \sin^2 \left(\dfrac{m\pi h}{2} \right) = 8h^{-2} + O(1).$

For ih or jh of size $O(1)$, say, $i > (m+1)/4$ or $j > (m+1)/4$, we have $\lambda_{i,j} = O(h^{-2})$, which is the same order as λ_{max}. Hence if the $((m+1)/4)^2$ smallest eigencomponents in the error could be annihilated by solving a smaller problem, using a partial step of the form (12.3), then the ratio of the largest to the smallest remaining eigenvalue would be $O(1)$, independent of h. The bound (12.20) on the convergence rate of iteration (12.3–12.4) with $G = \gamma I$ would be a constant less than one and independent of h!

Note also that the eigenvectors corresponding to the smaller values of i and j are "low frequency." That is, the sine functions do not go through many periods as p and q range from 1 to m. Thus these eigenvectors could be represented on a coarser grid. We now show how the annihilation of the small eigencomponents can be accomplished, approximately, by solving the problem on a coarser grid.

Assume that $m+1$ is even and let $\hat{m} = (m-1)/2$ be the number of interior points in each direction of a coarser grid with spacing $\hat{h} = 2h$. Let $\hat{n} = \hat{m}^2$. Define the coarse-to-fine prolongation matrix $I^n_{\hat{n}}$ to be *linear* interpolation along horizontal, vertical, and diagonal (southwest to northeast) lines. That is, if w is a vector defined at the nodes $(1,1) - (\hat{m},\hat{m})$ of the coarse grid, define

$$(I^n_{\hat{n}} w)_{p,q} = \begin{cases} w_{p/2,q/2} & \text{if } p,q \text{ even,} \\ \frac{1}{2}(w_{(p+1)/2,q/2} + w_{(p-1)/2,q/2}) & \text{if } p \text{ odd, } q \text{ even,} \\ \frac{1}{2}(w_{p/2,(q+1)/2} + w_{p/2,(q-1)/2}) & \text{if } p \text{ even, } q \text{ odd,} \\ \frac{1}{2}(w_{(p+1)/2,(q+1)/2} + w_{(p-1)/2,(q-1)/2}) & \text{if } p,q \text{ odd,} \end{cases}$$

(12.25) $p,q = 1,\ldots,m.$

THEOREM 12.1.2. *Let A be the 5-point Laplacian matrix so that $\lambda_{i,j}$ and $v^{(i,j)}$ satisfy (12.21–12.22), and let $I^n_{\hat{n}}$ be defined by (12.25). Let $v^{(1)},\ldots,v^{(s)}$ denote the eigenvectors corresponding to the s smallest eigenvalues, $\lambda_1 \leq \cdots \leq \lambda_s$. If v is any vector in $\mathrm{span}[v^{(1)},\ldots,v^{(s)}]$, with $\|v\| = 1$, then v can be written in the form*

(12.26) $v = A^{1/2} I^n_{\hat{n}} w + \delta, \quad \text{where } \|\delta\| \leq ch\lambda_s^{1/2}$

for some $\hat{n}$-vector w where $c = 2 + \sqrt{6}$.

Proof. First suppose that $v = v^{(i,j)}$ is an eigenvector. Then for any $\hat{n}$-vector w, we have

$$\|v^{(i,j)} - A^{1/2}I_{\hat{n}}^n w\| = \|A^{1/2}(\lambda_{i,j}^{-1/2}v^{(i,j)} - I_{\hat{n}}^n w)\| \le \|A^{1/2}\|\, \lambda_{i,j}^{-1/2}\, \|v^{(i,j)} - I_{\hat{n}}^n \tilde{w}\|,$$

where $\tilde{w} = \lambda_{i,j}^{1/2}w$. Since, from (12.24), the norm of $A^{1/2}$ is bounded by $2\sqrt{2}\,h^{-1}$, we have

$$(12.27) \qquad \|v^{(i,j)} - A^{1/2}I_{\hat{n}}^n w\| \le 2\sqrt{2}\,h^{-1}\lambda_{i,j}^{-1/2}\,\|v^{(i,j)} - I_{\hat{n}}^n \tilde{w}\|.$$

Let $\tilde{w}^{(i,j)}$ match $v^{(i,j)}$ at the nodes of the coarse grid so that

$$\left(I_{\hat{n}}^n \tilde{w}^{(i,j)}\right)_{p,q} = \begin{cases} v_{p,q}^{(i,j)} & \text{if } p, q \text{ even,} \\ \frac{1}{2}\left(v_{p-1,q}^{(i,j)} + v_{p+1,q}^{(i,j)}\right) & \text{if } p \text{ odd, } q \text{ even,} \\ \frac{1}{2}\left(v_{p,q-1}^{(i,j)} + v_{p,q+1}^{(i,j)}\right) & \text{if } p \text{ even, } q \text{ odd,} \\ \frac{1}{2}\left(v_{p-1,q-1}^{(i,j)} + v_{p+1,q+1}^{(i,j)}\right) & \text{if } p, q \text{ odd,} \end{cases}$$

for $p, q = 1, \ldots, m$. Then from formula (12.22) for $v^{(i,j)}$ it follows that

$$\left(v^{(i,j)} - I_{\hat{n}}^n \tilde{w}^{(i,j)}\right)_{p,q} = \begin{cases} 0 & \text{if } p, q \text{ even,} \\ v_{p,q}^{(i,j)}(1 - \cos(ih\pi)) & \text{if } p \text{ odd, } q \text{ even,} \\ v_{p,q}^{(i,j)}(1 - \cos(jh\pi)) & \text{if } p \text{ even, } q \text{ odd,} \\ v_{p,q}^{(i,j)}(1 - \cos(ih\pi)\cos(jh\pi)) - \\ \quad z_{p,q}^{(i,j)}\sin(ih\pi)\sin(jh\pi) & \text{if } p, q \text{ odd,} \end{cases}$$

$$(12.28) \qquad z_{p,q}^{(i,j)} \equiv 2h\cos(ihp\pi)\cos(jhq\pi) \text{ for } p, q \text{ odd.}$$

Note that if $z_{p,q}^{(i,j)}$ is defined by (12.28) for all points p and q, then the vectors $z^{(i,j)}$ are orthonormal (Exercise 12.1), as are the vectors $v^{(i,j)}$ (Exercise 9.3). If $z^{(i,j)}$ is defined to be 0 at the other grid points, then it can be checked (Exercise 12.1) that

$$\|z^{(i,j)}\|^2 \le 1/4 \quad \text{and} \quad \langle z^{(i,j)}, v^{(i,j)}\rangle = 0.$$

Summing over p and q and using the formula $1 - \cos x = 2\sin^2(x/2)$, we have

$$\begin{aligned} \|v^{(i,j)} - I_{\hat{n}}^n \tilde{w}^{(i,j)}\|^2 \;=\;& 4\sin^4(ih\pi/2) \sum_{p \text{ odd, } q \text{ even}} (v_{p,q}^{(i,j)})^2 \\ &+ 4\sin^4(jh\pi/2) \sum_{p \text{ even, } q \text{ odd}} (v_{p,q}^{(i,j)})^2 \\ &+ (1 - \cos(ih\pi)\cos(jh\pi))^2 \sum_{p,q \text{ odd}} (v_{p,q}^{(i,j)})^2 \\ &\sin^2(ih\pi)\sin^2(jh\pi) \sum_{p,q \text{ odd}} + (z_{p,q}^{(i,j)})^2. \end{aligned}$$

From (12.21) it follows that

$$(12.29) \qquad \max\{4\sin^4(ih\pi/2), 4\sin^4(jh\pi/2)\} \le \frac{1}{4}h^4\lambda_{i,j}^2,$$

$$
\begin{aligned}
(1 - \cos(ih\pi)\cos(jh\pi))^2 &= \left[1 - \left(1 - 2\sin^2(ih\pi/2)\right)\left(1 - 2\sin^2(jh\pi/2)\right)\right]^2 \\
&\le 4\left(\sin^2(ih\pi/2) + \sin^2(jh\pi/2)\right)^2 \\
&\le \frac{1}{4}h^4\lambda_{i,j}^2, \quad \text{and}
\end{aligned}
$$

$$(12.30)$$

$$
\begin{aligned}
\sin^2(ih\pi)\sin^2(jh\pi) &= [2\sin(ih\pi/2)\cos(ih\pi/2)]^2[2\sin(jh\pi/2)\cos(jh\pi/2)]^2 \\
&\le 16\sin^2(ih\pi/2)\sin^2(jh\pi/2) \\
&\le \frac{1}{2}h^4\lambda_{i,j}^2.
\end{aligned}
$$

$$(12.31)$$

Making these substitutions and using the fact that $\|v^{(i,j)}\|^2 = 1$ and $\|z^{(i,j)}\|^2 \le 1/4$, we can write

$$\|v^{(i,j)} - I_{\hat{n}}^n \tilde{w}^{(i,j)}\|^2 \le \frac{1}{4}h^4\lambda_{i,j}^2\|v^{(i,j)}\|^2 + \frac{1}{2}h^4\lambda_{i,j}^2\|z^{(i,j)}\|^2 \le \frac{3}{8}h^4\lambda_{i,j}^2$$

or

$$\|v^{(i,j)} - I_{\hat{n}}^n \tilde{w}^{(i,j)}\| \le \sqrt{\frac{3}{8}}h^2\lambda_{i,j}.$$

Combining this with (12.27) gives

$$(12.32) \qquad \|v^{(i,j)} - A^{1/2}I_{\hat{n}}^n w^{(i,j)}\| \le \sqrt{3}\, h\, \lambda_{i,j}^{1/2}.$$

Now let V_s be the matrix whose columns are the eigenvectors $v^{(i,j)}$ corresponding to the s smallest eigenvalues, and let $v = V_s\xi$ be an arbitrary vector in the span of the first s eigenvectors, with $\|v\| = \|\xi\| = 1$. Consider approximating v by the vector $A^{1/2}I_{\hat{n}}^n(\tilde{W}_s\Lambda_s^{-1/2}\xi)$, where $\tilde{W}_s$ has columns $\tilde{w}^{(i,j)}$ corresponding to the s smallest eigenvalues and Λ_s is the diagonal matrix of these eigenvalues. The difference $\delta \equiv v - A^{1/2}I_{\hat{n}}^n(\tilde{W}_s\Lambda_s^{-1/2}\xi)$ is given by

$$(12.33) \qquad \delta = A^{1/2}\,\triangle_s\,\Lambda_s^{-1/2}\xi,$$

where the columns of $\triangle_s$ are the vectors $\delta^{(i,j)} \equiv v^{(i,j)} - I_{\hat{n}}^n\tilde{w}^{(i,j)}$. From (12.28) the vector $\triangle_s\Lambda_s^{-1/2}\xi$ can be written in the form $d + e$, where

$$d = \begin{cases} 0 & \text{at even–even grid points,} \\ V_s^{oe}\hat{\xi} & \text{at odd–even grid points,} \\ V_s^{eo}\tilde{\xi} & \text{at even–odd grid points,} \\ V_s^{oo}\breve{\xi} & \text{at odd–odd grid points,} \end{cases}$$

$$e = \begin{cases} 0 & \text{at even–even, odd–even, and even–odd grid points,} \\ Z_s^{oo}\xi & \text{at odd–odd grid points,} \end{cases}$$

where V_s^{oe}, V_s^{eo}, V_s^{oo}, and Z_s^{oo} consist of the rows of V_s or Z_s (the matrix whose columns are the vectors $z^{(i,j)}$ defined in (12.28) for indices (i,j) corresponding to the s smallest eigenvalues) corresponding to the odd–even, even–odd, and odd–odd grid points and

$$\hat{\xi}_{i,j} = \lambda_{i,j}^{-1/2}\xi_{i,j}(1 - \cos(ih\pi)), \quad \tilde{\xi}_{i,j} = \lambda_{i,j}^{-1/2}\xi_{i,j}(1 - \cos(jh\pi)),$$

$$\breve{\xi}_{i,j} = \lambda_{i,j}^{-1/2}\xi_{i,j}(1 - \cos(ih\pi)\cos(jh\pi)), \quad \dot{\xi}_{i,j} = \lambda_{i,j}^{-1/2}\xi_{i,j}\sin(ih\pi)\sin(jh\pi).$$

Each of the matrices V_s^{oe}, V_s^{eo}, V_s^{oo}, and Z_s^{oo} has norm less than or equal to 1, because it is part of an orthogonal matrix. Since $\|\xi\| = 1$, it follows that

$$\begin{aligned} \|d\|^2 &\leq \max_{i,j}\lambda_{i,j}^{-1}(1 - \cos(ih\pi))^2 + \max_{i,j}\lambda_{i,j}^{-1}(1 - \cos(jh\pi))^2 \\ &\quad + \max_{i,j}\lambda_{i,j}^{-1}(1 - \cos(ih\pi)\cos(jh\pi))^2, \\ \|e\|^2 &\leq \max_{i,j}\lambda_{i,j}^{-1}\sin^2(ih\pi)\sin^2(jh\pi). \end{aligned}$$

Using (12.29–12.31), we have

$$\|d\|^2 \leq \frac{3}{4}h^4\lambda_s, \quad \|e\|^2 \leq \frac{1}{2}h^4\lambda_s,$$

and from (12.33) and (12.24), it follows that

$$\|\delta\| \leq \|A^{1/2}\|\,(\|d\| + \|e\|) \leq (2 + \sqrt{6})h\lambda_s^{1/2}. \quad \square$$

The constant $2 + \sqrt{6}$ in (12.26) is not the best possible estimate because the piecewise linear interpolant of $v^{(i,j)}$ used in the theorem is not the best possible coarse grid approximation to $v^{(i,j)}$.

COROLLARY 12.1.1. *If* $y \perp A^{1/2} \cdot \mathcal{R}(I_{\hat{n}}^n)$ *and* $\|y\| = 1$, *then*

$$(12.34) \qquad \sum_{i=1}^{s}\langle y, v^{(i)}\rangle^2 \leq c^2 h^2 \lambda_s, \quad c = 2 + \sqrt{6}.$$

Proof. The left-hand side of inequality (12.34) is the square of the norm of the vector $v = \sum_{i=1}^{s}\langle y, v^{(i)}\rangle v^{(i)}$, and according to Theorem 12.1.2 this vector satisfies

$$\frac{v}{\|v\|} = A^{1/2}I_{\hat{n}}^n w + \delta, \quad \|\delta\| \leq (2 + \sqrt{6})h\lambda_s^{1/2}$$

for some $\hat{n}$-vector w. The condition $y \perp A^{1/2}\mathcal{R}(I_{\hat{n}}^n)$ and $\|y\| = 1$ implies

$$\left|\left\langle y, \frac{v}{\|v\|}\right\rangle\right| = |\langle y, \delta\rangle| \leq \|\delta\|.$$

Since $\langle y, v\rangle = \langle v, v\rangle = \|v\|^2$, the desired result (12.34) follows. $\quad \square$

We now use Theorem 12.1.2 and Corollary 12.1.1 to bound the quantity on the right-hand side of (12.14), again assuming that $G = \gamma I$. In this case, inequality (12.14) can be written in the form

$$\sigma^2 \leq \max_{\substack{\|y\| = 1 \\ y \perp A^{1/2} \cdot \mathcal{R}(I_{\hat{n}}^n)}} \left(\sum_{i=1}^{s} \langle y, v^{(i)} \rangle^2 (1 - \gamma \lambda_i)^{2\ell} + \sum_{i=s+1}^{n} \langle y, v^{(i)} \rangle^2 (1 - \gamma \lambda_i)^{2\ell} \right).$$

Taking $\gamma = 2/(\lambda_n + \lambda_{s+1})$, we can write

$$\sigma^2 \leq \max_{\substack{\|y\| = 1 \\ y \perp A^{1/2} \cdot \mathcal{R}(I_{\hat{n}}^n)}} \left(\sum_{i=1}^{s} \langle y, v^{(i)} \rangle^2 + \left(\frac{\kappa' - 1}{\kappa' + 1} \right)^{2\ell} \left(1 - \sum_{i=1}^{s} \langle y, v^{(i)} \rangle^2 \right) \right),$$

where $\kappa' = \lambda_n / \lambda_{s+1}$. Applying Corollary 12.1.1 (and using the fact that a function of the form $x + ((\kappa' - 1)/(\kappa' + 1))^{2\ell}(1 - x)$ is an increasing function of x for $0 \leq x \leq 1$), this becomes

(12.35) $$\sigma^2 \leq c^2 h^2 \lambda_s + \left(\frac{\kappa' - 1}{\kappa' + 1} \right)^{2\ell} (1 - c^2 h^2 \lambda_s),$$

provided that $c^2 h^2 \lambda_s \leq 1$.

Now, from (12.24) we know that λ_n is bounded by αh^{-2}, where $\alpha = 8$. (We are using the symbolic constants c and α instead of their actual values because similar results hold for more general finite element and finite difference equations for some constants c and α that are independent of h but are not necessarily the same ones as for the model problem.) Let $\beta > 0$ be any number less than or equal to α and such that

$$c^2 \beta < 1.$$

Choose s so that λ_s is the largest eigenvalue of A less than or equal to βh^{-2}:

$$\lambda_s \leq \beta h^{-2}, \quad \lambda_{s+1} > \beta h^{-2}.$$

Then expression (12.35) becomes

$$\sigma^2 \leq c^2 \beta + \left(\frac{\kappa' - 1}{\kappa' + 1} \right)^{2\ell} (1 - c^2 \beta), \quad \kappa' \leq \frac{\alpha}{\beta}.$$

We thus obtain a bound on σ^2 that is *strictly less than one* and *independent of h*. For example, choosing $\beta = 1/(2c^2)$ gives

$$\sigma^2 \leq \frac{1}{2} + \frac{1}{2} \left(\frac{\kappa' - 1}{\kappa' + 1} \right)^{2\ell}, \quad \kappa' \leq 2\alpha c^2.$$

For the model problem, this establishes $\kappa' \leq 16(2 + \sqrt{6})^2$ and, for $\ell = 1$, $\sigma \leq .997$. This is a *large overestimate* of the actual contraction number for the two-grid method, but it does establish convergence at a rate that is independent of h. To obtain a better estimate of σ, it would be necessary to derive a sharper bound on the constant c in Theorem 12.1.2.

12.1.3. Extension to More General Finite Element Equations.
The key to the analysis of section 12.1.2 was Theorem 12.1.2, showing that
vectors in the span of eigenvectors associated with small eigenvalues of A
can be well approximated on a coarser grid. This is true in general for
standard finite element matrices and often for finite difference matrices. It is
a consequence of the fact that the (functions represented by the) eigenvectors
corresponding to smaller eigenvalues on both the fine and coarse grids provide
good approximations to eigenfunctions of the elliptic differential operator, and
hence they also approximate each other. For the analogues of Theorem 12.1.2
and Corollary 12.1.1 in a more general setting, see [64]. Similar results can be
found in [104].

12.1.4. Multigrid Methods. The two-grid algorithm described in sections 12.1.1–12.1.2 is not practical in most cases because it requires solving a
linear system on a grid of spacing $2h$. Usually this is still too large a problem to
solve directly. The algorithm could be applied recursively, and the solution to
the problem on the coarser grid could be obtained by projecting the right-hand
side onto a still coarser grid, solving a linear system there, interpolating the
solution back to the finer grid, performing relaxation steps there, and repeating
this cycle until convergence. If the grid-$2h$ problem is solved very accurately,
however, this method would also be time consuming. A number of cycles might
be required to solve the problems on coarser grids before ever returning to the
fine grid where the solution is actually needed.

Instead, coarser grid problems can be "solved" very inaccurately by
performing just one relaxation sweep until the coarsest level is reached, at
which point the problem is solved directly. Let grid levels $0, 1, \ldots, J$ be defined
with maximum mesh spacings $h_0 \leq h_1 \leq \cdots \leq h_J$, and let $A^{(j)}$ denote the
coefficient matrix for the problem at level j. The linear system on the finest
grid is $Au = f$, where $A \equiv A^{(0)}$. The multigrid *V-cycle* consists of the following
steps.

> Given an initial guess u_0, compute $r_0 \equiv r_0^{(0)} = f - Au_0$. For
> $k = 1, 2, \ldots,$
>
> > For $j = 1, \ldots, J - 1$,
> > Project $r_{k-1}^{(j-1)}$ onto grid level j; that is, set
> >
> > $$f^{(j)} = I_{j-1}^{j} r_{k-1}^{(j-1)},$$
> >
> > where I_{j-1}^{j} is the restriction matrix from grid level $j - 1$
> > to grid level j.
> > Perform a relaxation sweep (with zero initial guess) on
> > grid level j; that is, solve
> >
> > $$G\delta_{k-1}^{(j)} = f^{(j)}$$

and compute

$$r_{k-1}^{(j)} = f^{(j)} - A^{(j)}\delta_{k-1}^{(j)}.$$

endfor

Project $r_{k-1}^{(J-1)}$ onto grid level J by setting $f^{(J)} = I_{J-1}^{J} r_{k-1}^{(J-1)}$ and solve on the coarsest grid $A^{(J)}d_{k-1}^{(J)} = f^{(J)}$.

For $j = J - 1, \ldots, 1$,

Interpolate $d_{k-1}^{(j+1)}$ to grid level j and add to $\delta_{k-1}^{(j)}$; that is, replace

$$\delta_{k-1}^{(j)} \leftarrow \delta_{k-1}^{(j)} + I_{j+1}^{j}d_{k-1}^{(j+1)},$$

where I_{j+1}^{j} is the prolongation matrix from grid level $j+1$ to grid level j.

Perform a relaxation sweep with initial guess $\delta_{k-1}^{(j)}$ on grid level j; that is, set

$$d_{k-1}^{(j)} = \delta_{k-1}^{(j)} + G^{-1}(f^{(j)} - A^{(j)}\delta_{k-1}^{(j)}).$$

endfor

Interpolate $d_{k-1}^{(1)}$ to grid level 0 and replace $u_{k-1} \leftarrow u_{k-1} + I_1^0 d_{k-1}^{(1)}$.

Perform a relaxation sweep with initial guess u_{k-1} on grid level 0; that is, set $u_k = u_{k-1}+G^{-1}(f - Au_{k-1})$. Compute the new residual $r_k \equiv r_k^{(0)} = f - Au_k$.

This iteration is called a *V-cycle* because it consists of going down through the grids from fine to coarse, performing a relaxation sweep on each grid, then coming back up from coarse to fine, and again performing a relaxation sweep at each level, as pictured below. (Sometimes an initial relaxation sweep on the fine grid is performed before projecting the residual onto the next coarser grid.) Other patterns of visiting the grids are also possible. In the *W-cycle*, for instance, one uses two V-cycles at each of the coarser levels, resulting in a pattern like that shown below for four grid levels. In the *full multigrid V-cycle*, one starts on the coarsest grid, goes up one level and then back down, up two levels and then back down, etc., until the finest level is reached. This provides the initial guess for the standard V-cycle, which is then performed.

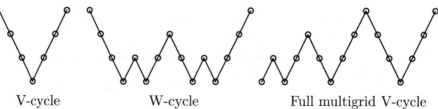

V-cycle   W-cycle   Full multigrid V-cycle

The restriction and prolongation matrices I_j^{j+1} and I_{j+1}^j, as well as the relaxation scheme with matrix G, can be tuned to the particular problem. For the model problem, the linear interpolation matrix I_{j+1}^j is appropriate, although it is not the only choice, and it is reasonable to define the restriction matrix I_j^{j+1} to be $I_{j+1}^{j^T}$, as in section 12.1.1. The damped Jacobi relaxation scheme described in section 12.1.1 is convenient for analysis, but other relaxation schemes may perform better in practice. The red–black Gauss–Seidel relaxation method is often used. (That is, if nodes are ordered so that the matrix A has the form (9.9), then G is taken to be the lower triangle of A.)

Figure 12.1 shows the convergence of the multigrid V-cycle with red–black Gauss–Seidel relaxation for the model problem for grid sizes $h = 1/64$ and $h = 1/128$. The coarsest grid, on which the problem was solved directly, was of size $h = 1/4$. Also shown in Figure 12.1 is the convergence curve for MICCG(0). The work per iteration (or per cycle for multigrid) for these two algorithms is similar. During a multigrid V-cycle, a (red–black) Gauss–Seidel relaxation step is performed once on the fine grid and twice on each of the coarser grids. Since the number of points on each coarser level grid is about $1/4$ that of the finer grid, this is the equivalent of about 1 and 2/3 Gauss–Seidel sweeps on the finest grid. After the fine grid relaxation is complete, a new residual must be computed, requiring an additional matrix–vector multiplication on the fine grid. In the MICCG(0) algorithm, backsolving with the L and L^T factors of the MIC decomposition is twice the work of backsolving with a single lower triangular matrix in the Gauss–Seidel method, but only one matrix–vector multiplication is performed at each step. The CG algorithm also requires some inner products that are not present in the multigrid algorithm, but the multigrid method requires prolongation and restriction operations that roughly balance with the work for the inner products. The exact operation count is implementation dependent, but for the implementation used here (which was designed for a general 5-point matrix, not just the Laplacian), the operation count per cycle/iteration was about $41n$ for multigrid and about $30n$ for MICCG(0).

It is clear from Figure 12.1 that for the model problem, the multigrid method is by far the most efficient of the iterative methods we have discussed. Moreover, the multigrid method demonstrated here is *not* the best. The number of cycles can be reduced even further (from 9 down to about 5 to achieve an error of size 10^{-6}) by using the W-cycle or the full multigrid V-cycle, with more accurate restriction and prolongation operators. The work per cycle is somewhat greater, but the reduction in number of cycles more than makes up for the slight increase in cycle time. (See Exercise 12.2.)

The multigrid method described here works well for a variety of problems, including nonsymmetric differential equations, such as $-\triangle u + c u_x = f$, as well as for the model problem. It should be noted, however, that while the performance of ICCG and MICCG is not greatly changed if the model problem

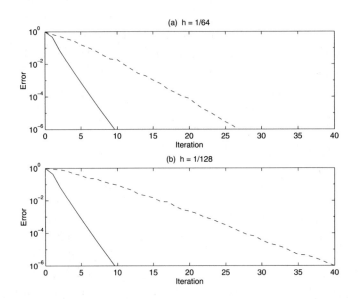

FIG. 12.1. *Convergence of the multigrid V-cycle (solid) and MICCG(0) (dashed) for the model problem.*

is replaced by the diffusion equation (9.1) with a highly varying diffusion coefficient, this is not the case for the multigrid method. The multigrid algorithm used for the model problem will still converge at a rate that is independent of h if applied to the diffusion equation, but the convergence rate will be *greatly* affected by the variation in the diffusion coefficient. For problems with discontinuous diffusion coefficients, linear interpolation, as used here, is not really appropriate. It should be replaced by a form of interpolation that takes account of the discontinuities [1].

For this reason, instead of thinking of *the* multigrid method, one should view the multigrid approach as a framework for developing iterative methods (that is, preconditioners to be used with simple iteration or other Krylov subspace methods). Sometimes, based on known properties of the differential equation, one can identify appropriate prolongation, restriction, and relaxation matrices that will result in a multigrid method whose convergence rate is not only independent of h but is much better than that of other methods for realistic mesh sizes. One should look for both a relaxation method that damps high frequencies very rapidly and restriction and prolongation matrices having the property that the low frequency components of the error are greatly reduced when the residual is projected onto a coarser grid, a problem is solved on that grid, and the solution is interpolated to the finer grid and added to the previous approximation. Such multigrid methods have been developed for a wide variety of physical problems. This is not always possible, however. For problems that are barely resolved on the grid of interest, it may be unclear how the problem should even be defined on coarser level grids, and one cannot expect to gain much information from a "solution" on a such a grid.

12.1.5. Multigrid as a Preconditioner for Krylov Subspace Methods.
Some multigrid aficionados will argue that if one has used the proper restriction, prolongation, and relaxation operators, then the multigrid algorithm will require so few cycles (one or two full multigrid V-cycles to reach the level of truncation error) that it is almost pointless to try to accelerate it with CG-like methods. This may be true, but unfortunately such restriction, prolongation, and relaxation schemes are not always known. In such cases, CG, GMRES, QMR, or BiCGSTAB acceleration may help.

Equivalently, one can consider multigrid as a preconditioner for one of these Krylov subspace methods. To solve an equation $Mz = r$ with the multigrid V-cycle preconditioner M as coefficient matrix, one simply performs one multigrid V-cycle with right-hand side r and initial guess zero.

For some interesting examples using multigrid as a preconditioner for GMRES and BiCGSTAB, see [108]. The use of multigrid (with damped Jacobi relaxation) as a preconditioner for the CG algorithm for solving diffusion-like equations is described in [4].

12.2. Basic Ideas of Domain Decomposition Methods.
Simulation problems often involve complicated structures such as airplanes and automobiles. Limitations on computer time and storage may prevent the modeling of the entire structure at once, so instead a piece of the problem is studied, e.g., an airplane wing. If different parts of the problem could be solved independently and then the results somehow pieced together to give the solution to the entire problem, then a loosely coupled array of parallel processors could be used for the task. This is one of the motivations for domain decomposition methods. Even if the domain of the problem is not so complicated, one might be able to break the domain into pieces on which the problem is more easily solved, e.g., rectangles on which a fast Poisson solver could be used or subdomains more suitable for multigrid methods. If the solutions of the subproblems can be combined in a clever way to solve the overall problem, then this may provide a faster and more parallel solution method than applying a standard iterative method directly to the large problem. We will see that this solution approach is equivalent to using a preconditioner that involves solving on subdomains. The clever way of combining the solutions from subdomains is usually a CG-like iterative method.

Domain decomposition methods fall roughly into two classes—those using *overlapping domains*, such as the additive and multiplicative Schwarz methods, and those using *nonoverlapping domains*, which are sometimes called *substructuring* methods. If one takes a more general view of the term "subdomain," then the subdomains need not be contiguous parts of the physical domain at all but may be parts of the solution space, such as components that can be represented on a coarser grid and those that cannot. With this interpretation, multigrid methods fall under the heading of domain decomposition methods.

In this chapter, we describe some basic domain decomposition methods but give little of the convergence theory. For further discussion, see [123] or [77].

12.2.1. Alternating Schwarz Method.

Let $\mathcal{L}$ be a differential operator defined on a domain Ω, and suppose we wish to solve the boundary value problem

$$\mathcal{L}u = f \quad \text{in } \Omega,$$

(12.36)
$$u = g \quad \text{on } \partial\Omega.$$

The domain Ω is an open set in the plane or in 3-space, and $\partial\Omega$ denotes the boundary of Ω. We denote the closure of Ω by $\bar{\Omega} \equiv \Omega \cup \partial\Omega$. We have chosen Dirichlet boundary conditions ($u = g$ on $\partial\Omega$), but Neumann or Robin boundary conditions could be specified as well.

The domain Ω might be divided into two overlapping pieces, Ω_1 and Ω_2, such that $\Omega = \Omega_1 \cup \Omega_2$, as pictured in Figure 12.2. Let Γ_1 and Γ_2 denote the parts of the boundaries of Ω_1 and Ω_2, respectively, that are not part of the boundary of Ω. To solve this problem, one might guess the solution on Γ_1 and solve the problem

$$\mathcal{L}u_1 = f \quad \text{in } \Omega_1,$$

(12.37)
$$u_1 = g \quad \text{on } \partial\Omega \cap \partial\Omega_1, \quad u_1 = g_1 \quad \text{on } \Gamma_1,$$

where g_1 is the initial guess for the solution on Γ_1. Letting g_2 be the value of u_1 on Γ_2, one then solves

$$\mathcal{L}u_2 = f \quad \text{in } \Omega_2,$$

(12.38)
$$u_2 = g \quad \text{on } \partial\Omega \cap \partial\Omega_2, \quad u_2 = g_2 \quad \text{on } \Gamma_2.$$

If the computed solutions u_1 and u_2 are the same in the region where they overlap then the solution to problem (12.36) is

$$u = \begin{cases} u_1 & \text{in } \Omega_1 \backslash \Omega_2, \\ u_2 & \text{in } \Omega_2 \backslash \Omega_1, \\ u_1 \equiv u_2 & \text{in } \Omega_1 \cap \Omega_2. \end{cases}$$

If the values of u_1 and u_2 differ in the overlap region, then the process can be repeated, replacing g_1 by the value of u_2 on Γ_1, and re-solving problem (12.37), etc. This idea was introduced by Schwarz in 1870 [120], not as a computational technique, but to establish the existence of solutions to elliptic problems on regions where analytic solutions were not known. When used as a computational technique it is called the *alternating Schwarz method*.

A slight variation of the alternating Schwarz method, known as the *multiplicative Schwarz method*, is more often used in computations. Let the problem (12.36) be discretized using a standard finite difference or finite element method, and assume that the overlap region is sufficiently wide so

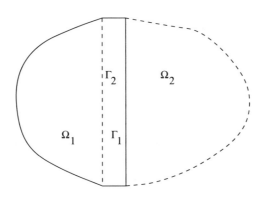

FIG. 12.2. *Decomposition of domain into two overlapping pieces.*

that nodes in $\Omega_1 \backslash \Omega_2$ do not couple to nodes in $\Omega_2 \backslash \Omega_1$, and vice versa. Assume also that the boundaries Γ_1 and Γ_2 are grid lines. If nodes in $\Omega_1 \backslash \Omega_2$ are numbered first, followed by nodes in $\Omega_1 \cap \Omega_2$, and then followed by nodes in $\Omega_2 \backslash \Omega_1$, then the discretized problem can be written in the form

$$(12.39) \qquad \begin{pmatrix} A_{11} & A_{12} & 0 \\ A_{21} & A_{22} & A_{23} \\ 0 & A_{32} & A_{33} \end{pmatrix} \begin{pmatrix} u_{\Omega_1 \backslash \Omega_2} \\ u_{\Omega_1 \cap \Omega_2} \\ u_{\Omega_2 \backslash \Omega_1} \end{pmatrix} = \begin{pmatrix} f_{\Omega_1 \backslash \Omega_2} \\ f_{\Omega_1 \cap \Omega_2} \\ f_{\Omega_2 \backslash \Omega_1} \end{pmatrix},$$

where the right-hand side vector f includes contributions from the boundary term $u = g$ on $\partial \Omega$.

Starting with an initial guess $u^{(0)}$ (which actually need only be defined on Γ_1 for a standard 5-point discretization or, more generally, on points in $\Omega_2 \backslash \Omega_1$ that couple to points in $\Omega_1 \cap \Omega_2$), the multiplicative Schwarz method for the discretized system generates approximations $u^{(k)}$, $k = 1, 2, \ldots$, satisfying

$$\begin{pmatrix} A_{11} & A_{12} \\ A_{21} & A_{22} \end{pmatrix} \begin{pmatrix} u_{\Omega_1 \backslash \Omega_2} \\ u_{\Omega_1 \cap \Omega_2} \end{pmatrix}^{(k)} = \begin{pmatrix} f_{\Omega_1 \backslash \Omega_2} \\ f_{\Omega_1 \cap \Omega_2} - A_{23} u_{\Omega_2 \backslash \Omega_1}^{(k-1)} \end{pmatrix},$$

$$(12.40) \qquad \begin{pmatrix} A_{22} & A_{23} \\ A_{32} & A_{33} \end{pmatrix} \begin{pmatrix} u_{\Omega_1 \cap \Omega_2} \\ u_{\Omega_2 \backslash \Omega_1} \end{pmatrix}^{(k)} = \begin{pmatrix} f_{\Omega_1 \cap \Omega_2} - A_{21} u_{\Omega_1 \backslash \Omega_2}^{(k)} \\ f_{\Omega_2 \backslash \Omega_1} \end{pmatrix}.$$

The first equation corresponds to solving the problem on Ω_1, using boundary data obtained from $u_{\Omega_2 \backslash \Omega_1}^{(k-1)}$. The second equation corresponds to solving the problem on Ω_2, using boundary data obtained from $u_{\Omega_1 \backslash \Omega_2}^{(k)}$.

Note that this is somewhat like a *block Gauss–Seidel method* for (12.39), since one solves the first block equation using old data on the right-hand side and the second block equation using updated data on the right-hand side, but in this case the blocks overlap:

$$\begin{pmatrix} A_{11} & A_{12} & | & 0 \\ - & - & - & - \\ A_{21} & | & A_{22} & | & A_{23} \\ - & - & - & - \\ 0 & | & A_{32} & A_{33} \end{pmatrix}.$$

Let E_i^T, $i = 1, 2$, be the rectangular matrix that takes a vector defined on all of Ω and *restricts* it to Ω_i:

$$E_1^T \begin{pmatrix} v_{\Omega_1 \setminus \Omega_2} \\ v_{\Omega_1 \cap \Omega_2} \\ v_{\Omega_2 \setminus \Omega_1} \end{pmatrix} = \begin{pmatrix} v_{\Omega_1 \setminus \Omega_2} \\ v_{\Omega_1 \cap \Omega_2} \end{pmatrix}, \quad E_2^T \begin{pmatrix} v_{\Omega_1 \setminus \Omega_2} \\ v_{\Omega_1 \cap \Omega_2} \\ v_{\Omega_2 \setminus \Omega_1} \end{pmatrix} = \begin{pmatrix} v_{\Omega_1 \cap \Omega_2} \\ v_{\Omega_2 \setminus \Omega_1} \end{pmatrix}.$$

The matrix E_i takes a vector defined on Ω_i and *extends* it with zeros to the rest of Ω. The matrices on the left in (12.40) are of the form $E_i^T A E_i$, where A is the coefficient matrix in (12.39). Using this notation, iteration (12.40) can be written in the equivalent form

$$E_1^T u^{(k)} \leftarrow E_1^T u^{(k-1)} + (E_1^T A E_1)^{-1} E_1^T (f - A u^{(k-1)}), \quad u_{\Omega_2 \setminus \Omega_1}^{(k)} \leftarrow u_{\Omega_2 \setminus \Omega_1}^{(k-1)},$$

$$E_2^T u^{(k)} \leftarrow E_2^T u^{(k)} + (E_2^T A E_2)^{-1} E_2 (f - A u^{(k)}).$$

Writing this as two half-steps and extending the equations to the entire domain, the iteration becomes

$$u^{(k-1/2)} = u^{(k-1)} + E_1 (E_1^T A E_1)^{-1} E_1^T (f - A u^{(k-1)}),$$

$$u^{(k)} = u^{(k-1/2)} + E_2 (E_2^T A E_2)^{-1} E_2^T (f - A u^{(k-1/2)}).$$

Defining $B_i \equiv E_i (E_i^T A E_i)^{-1} E_i^T$, these two half-steps can be combined to give

(12.41)　　　$u^{(k)} = u^{(k-1)} + (B_1 + B_2 - B_2 A B_1)(f - A u^{(k-1)}).$

This is the simple iteration method described in section 2.1 with preconditioner $M^{-1} = B_1 + B_2 - B_2 A B_1$.

One could also consider solving (12.39) using an *overlapping block Jacobi-type method*; that is, using data from the previous iterate in the right-hand sides of both equations in (12.40). This leads to the set of equations

$$u_{\Omega_1}^{(k)} = u_{\Omega_1}^{(k-1)} + (E_1^T A E_1)^{-1} E_1^T (f - A u^{(k-1)}),$$

$$u_{\Omega_2}^{(k)} = u_{\Omega_2}^{(k-1)} + (E_2^T A E_2)^{-1} E_2^T (f - A u^{(k-1)}),$$

where $u_{\Omega_i}^{(k)} \equiv E_i^T u^{(k)}$. The value of $u^{(k)}$ in the overlap region has been set in two different ways by these equations. For the multiplicative Schwarz method, we used the second equation to define the value of $u^{(k)}$ in the overlap region; for this variant it is customary to take $u_{\Omega_1 \cap \Omega_2}^{(k)}$ to be the *sum* of the two values defined by these equations. This leads to the *additive Schwarz method*:

$$(12.42) \qquad u^{(k)} = u^{(k-1)} + (B_1 + B_2)(f - Au^{(k-1)}).$$

In this case, the preconditioner $M^{-1} = B_1 + B_2$ is Hermitian if A is Hermitian, so the simple iteration (12.42) can be replaced by the CG or MINRES algorithm. In fact, some form of acceleration or damping factor must be used with iteration (12.42) to ensure convergence. To solve the preconditioning equation $Mz = r$, one simply solves a problem on each of the two subdomains independently, using boundary data from the previous iterate, and adds the results.

We will not prove any convergence results for the additive and multiplicative Schwarz preconditioners, but note that for the model problem and many other elliptic differential equations using say, GMRES acceleration, these preconditioners have the following properties (see [123]):

- The number of iterations (to reduce the initial residual norm by a fixed factor) is independent of the mesh size, provided that the overlap region is kept fixed. (Remember, however, that with only two subdomains the solution time on each subdomain grows as the mesh size is decreased!)

- The overlap region can be quite small without greatly affecting the convergence rate.

- The number of iterations for the multiplicative variant is about half that for the additive algorithm. (This is similar to the relation between ordinary Gauss–Seidel and Jacobi iterations (10.20, 10.22).)

12.2.2. Many Subdomains and the Use of Coarse Grids. The multiplicative and additive Schwarz methods are easily extended to multiple subdomains. We will concentrate on the additive variant because it provides greater potential for parallelism. If the region Ω is divided into J overlapping subregions $\Omega_1, \ldots, \Omega_J$, then the additive Schwarz preconditioner is

$$(12.43) \qquad M^{-1} \equiv \sum_{i=1}^{J} B_i,$$

where $B_i = E_i(E_i^T A E_i)^{-1} E_i^T$, as in the previous section. To apply M^{-1} to a vector r, one solves a problem on each subdomain with right-hand side $E_i^T r$ and adds the results. These subdomain solves can be carried out in parallel, so it is desirable to have a large number of subdomains.

Consider the Krylov space generated by f and $M^{-1}A$:

$$\text{span}\{f, (M^{-1}A)f, (M^{-1}A)^2 f, \ldots\}.$$

(For a Hermitian positive definite problem we could equally well consider the Krylov space generated by $L^{-H}f$ and $L^{-1}AL^{-H}$, where $M = LL^H$.) Suppose f has nonzero components in only one of the subdomains, say, Ω_1. Since a standard finite difference or finite element matrix A contains only local couplings, the vector Af will be nonzero only in subdomains that overlap with Ω_1 (or, perhaps, subdomains that are separated from Ω_1 by just a few mesh widths). It is only for these subdomains that the right-hand side $E_i^T f$ of the subdomain problem will be nonzero and hence that $M^{-1}Af$ will be nonzero. If this set of subdomains is denoted S_1, then it is only for subdomains that overlap with (or are separated by just a few mesh widths from) subdomains in S_1 that the next Krylov vector $(M^{-1}A)^2 f$ will be nonzero. And so on. The number of Krylov space vectors will have to reach the length of the shortest path from Ω_1 to the most distant subregion, say, Ω_J, before any of the Krylov vectors will have nonzero components in that subregion. Yet the solution $u(\mathbf{x})$ of the differential equation and the vector u satisfying the discretized problem $Au = f$ may well have nonzero (and not particularly small) components in all of the subdomains. Hence any Krylov space method for solving $Au = f$ with a zero initial guess and the additive Schwarz preconditioner will require at least this shortest path length number of iterations to converge (that is, to satisfy a reasonable error tolerance). As the number of subdomains increases, the shortest path between the most distant subregions also increases, so the number of iterations required by, say, the GMRES method with the additive Schwarz preconditioner will also increase. The reason is that there is no mechanism for global communication among the subdomains.

An interesting cure for this problem was proposed by Dryja and Widlund [36]. In addition to the subdomain solves, solve the problem on a coarse grid whose elements are the subregions of the original grid. If this problem is denoted $A_C u_C = f_C$ and if I_C^F denotes an appropriate type of interpolation (say, linear interpolation) from the coarse to the fine grid, then the preconditioner M^{-1} in (12.43) is replaced by

$$(12.44) \qquad M^{-1} = \sum_{i=1}^{J} B_i + I_C^F A_C^{-1} (I_C^F)^T.$$

It turns out that this small amount of global communication is sufficient to eliminate the dependence of the number of iterations on the number of subdomains. For this *two-level method*, the number of iterations is independent of both the mesh width h and the subdomain size H, assuming that the size of the overlap region is $O(H)$.

A two-level method such as this, however, must still require more than $O(n)$ work if both the subdomain and coarse grid solvers require more than $O(p)$ work, where p is the number of points in the subdomain or coarse grid. With just a few large subdomains, the subdomain solves will be too expensive and with many small subdomains, the coarse grid solve will be too expensive. For this reason, *multilevel algorithms* were developed to solve the large subproblems recursively by another application of the two-level preconditioner. The multilevel domain decomposition methods bear much resemblance to (and are sometimes identical with) standard multigrid methods described in section 12.1.

12.2.3. Nonoverlapping Subdomains.

Many finite element codes use a decomposition of the domain into nonoverlapping subregions to define an ordering of the unknowns for use with Gaussian elimination. If the domain Ω is divided into two pieces, Ω_1 and Ω_2, with interface Γ, and if points in the interior of Ω_1 (that do not couple to points in Ω_2) are numbered first, followed by points in the interior of Ω_2 (that do not couple to points in Ω_1) and then by points on the interface Γ, then the linear system takes the form

(12.45)
$$\begin{pmatrix} A_{11} & 0 & A_{1,\Gamma} \\ 0 & A_{22} & A_{2,\Gamma} \\ A_{\Gamma,1} & A_{\Gamma,2} & A_{\Gamma,\Gamma} \end{pmatrix} \begin{pmatrix} u_1 \\ u_2 \\ u_\Gamma \end{pmatrix} = \begin{pmatrix} f_1 \\ f_2 \\ f_\Gamma \end{pmatrix}.$$

If the matrices A_{11} and A_{22} can be inverted easily, then the variables u_1 and u_2 can be eliminated using Gaussian elimination and a much smaller *Schur complement* problem solved on the interface Γ:

$$(A_{\Gamma,\Gamma} - A_{\Gamma,1}A_{11}^{-1}A_{1,\Gamma} - A_{\Gamma,2}A_{22}^{-1}A_{2,\Gamma})u_\Gamma =$$

(12.46)
$$f_\Gamma - A_{\Gamma,1}A_{11}^{-1}f_1 - A_{\Gamma,2}A_{22}^{-1}f_2.$$

Once u_Γ is known, it can be substituted into (12.45) to obtain u_1 and u_2.

The coefficient matrix in (12.46) is small but dense and very expensive to form. It can be applied to a vector, however, by performing a few sparse matrix vector multiplications and solving on the subdomains. Hence an iterative method might be applied to (12.46). This is the idea of *iterative substructuring* methods. It is an idea that we have already seen in the solution of the transport equation in section 9.2. The source iteration described there can be thought of as an angular domain decomposition method using iterative substructuring (although the source iteration method was actually developed before these terms came into widespread use).

As we saw in section 9.2, the simple iterative substructuring method for that problem was equivalent to a block Gauss–Seidel iteration for the original linear system. While the transport equation is not elliptic, if the same idea were applied to a linear system arising from an elliptic differential equation, the convergence rate would not be independent of the mesh size. To obtain a convergence rate that is independent of, or only weakly dependent on, the mesh size, a preconditioner is needed for the system (12.46). Since the actual matrix in (12.46) is never formed, the standard Jacobi, Gauss–Seidel, SOR, and incomplete Cholesky-type preconditioners cannot be used.

Many preconditioners for (12.46) have been proposed, and a survey of these preconditioners is beyond the scope of this book. For a discussion of several such *interface preconditioners*, see [123].

Comments and Additional References.

Multigrid methods were first introduced by Fedorenko [49], and an early analysis was given by Bakhvalov [9]. A seminal paper by Brandt demonstrated the effectiveness of these techniques for a variety of problems [19]. An excellent introduction to multigrid methods, without much formal analysis, is given in a short book by Briggs [21], and additional information can be found in [18, 77, 78, 84, 98, 141, 146].

Exercises.

12.1. Show that the vectors $z^{(i,j)}$, $i, j = 1, \ldots, m$ with components

$$z_{p,q}^{(i,j)} = 2h \cos(ihp\pi) \cos(jhq\pi), \quad p, q = 1, \ldots, m,$$

defined on an m-by-m grid with spacing $h = 1/(m+1)$, form an orthonormal set. Show that if $\hat{z}_{p,q}^{(i,j)}$ is equal to $z_{p,q}^{(i,j)}$ if p and q are both odd but 0 otherwise, then

$$\|\hat{z}^{(i,j)}\| \leq \frac{1}{2} \quad \text{and} \quad \langle \hat{z}^{(i,j)}, v^{(i,j)} \rangle = 0,$$

where $v^{(i,j)}$ is defined in (12.22).

12.2. A *work unit* is often defined as the number of operations needed to perform one Gauss–Seidel sweep on the finest grid in a multigrid method. For a two-dimensional problem, approximately how many work units are required by (a) a multigrid V-cycle with a presmoothing step (that is, with a relaxation sweep performed on the fine grid at the beginning of the cycle as well as at the end), (b) a multigrid W-cycle, and (c) a full multigrid V-cycle? (You can ignore the work for restrictions and prolongations.) How do your answers change for a three-dimensional problem, assuming that each coarser grid still has mesh spacing equal to twice that of the next finer grid?

12.3. What is the preconditioner M in the iteration (12.5) if $\ell = 1$? Compute the two-grid preconditioner M described in section 12.1.2 for a small model problem. Is it a regular splitting? Are the entries of M close to those of A? (It is unlikely that one would choose this matrix as a preconditioner for A if one looked only at the entries of A and did not consider the origin of the problem!)

12.4. In the two-level additive Schwarz method, suppose that each subdomain and coarse grid solve requires $O(p^{3/2})$ work, where p is the number of points in the subdomain or coarse grid. (This is the cost of backsolving with a banded triangular factor for a matrix of order p with bandwidth $p^{1/2}$.) Assuming that the number of points on the coarse grid is approximately equal to the number of subdomains, what is the optimal number of subdomains needed to minimize the total work in applying the preconditioner (12.44), and how much total work is required?

References

[1] R. Alcouffe, A. Brandt, J. Dendy, and J. Painter, *The multigrid method for diffusion equations with strongly discontinuous coefficients*, SIAM J. Sci. Statist. Comput., 2 (1981), pp. 430–454.

[2] M. Arioli and C. Fassino, *Roundoff error analysis of algorithms based on Krylov subspace methods*, BIT, 36 (1996), pp. 189–205.

[3] S. F. Ashby, P. N. Brown, M. R. Dorr, and A. C. Hindmarsh, *A linear algebraic analysis of diffusion synthetic acceleration for the Boltzmann transport equation*, SIAM J. Numer. Anal., 32 (1995), pp. 179–214.

[4] S. F. Ashby, R. D. Falgout, S. G. Smith, and T. W. Fogwell, *Multigrid preconditioned conjugate gradients for the numerical simulation of groundwater flow on the Cray T3D*, American Nuclear Society Proceedings, Portland, OR, 1995.

[5] S. F. Ashby, T. A. Manteuffel, and P. E. Saylor, *A taxonomy for conjugate gradient methods*, SIAM J. Numer. Anal., 27 (1990), pp. 1542–1568.

[6] O. Axelsson, *A generalized SSOR method*, BIT, 13 (1972), pp. 442–467.

[7] O. Axelsson, *Bounds of eigenvalues of preconditioned matrices*, SIAM J. Matrix Anal. Appl., 13 (1992), pp. 847–862.

[8] O. Axelsson and H. Lu, *On eigenvalue estimates for block incomplete factorization methods*, SIAM J. Matrix Anal. Appl., 16 (1995), pp. 1074–1085.

[9] N. S. Bakhvalov, *On the convergence of a relaxation method with natural constraints on the elliptic operator*, U.S.S.R. Comput. Math. and Math. Phys., 6 (1966), pp. 101–135.

[10] R. Barrett, M. Berry, T. F. Chan, J. Demmel, J. Donato, J. Dongarra, V. Eijkhout, R. Pozo, C. Romine, and H. van der Vorst, *Templates for the Solution of Linear Systems: Building Blocks for Iterative Methods*, SIAM, Philadelphia, PA, 1995.

[11] T. Barth and T. Manteuffel, *Variable metric conjugate gradient methods*, in PCG '94: Matrix Analysis and Parallel Computing, M. Natori and T. Nodera, eds., Yokohama, 1994.

[12] T. Barth and T. Manteuffel, *Conjugate gradient algorithms using multiple recursions*, in Linear and Nonlinear Conjugate Gradient-Related Methods, L. Adams and J. L. Nazareth, eds., SIAM, Philadelphia, PA, 1996.

[13] R. Beauwens, *Approximate factorizations with S/P consistently ordered M-factors*, BIT, 29 (1989), pp. 658–681.

[14] R. Beauwens, *Modified incomplete factorization strategies*, in Preconditioned Conjugate Gradient Methods, O. Axelsson and L. Kolotilina, eds., Lecture Notes in Mathematics 1457, Springer-Verlag, Berlin, New York, 1990, pp. 1–16.

[15] M. W. Benson and P. O. Frederickson, *Iterative solution of large sparse systems arising in certain multidimensional approximation problems*, Utilitas Math., 22 (1982), pp. 127–140.

[16] M. Benzi, C. D. Meyer, and M. Tůma, *A sparse approximate inverse preconditioner for the conjugate gradient method*, SIAM J. Sci. Comput., 17 (1996), pp. 1135–1149.

[17] M. Benzi and M. Tůma, *A sparse approximate inverse preconditioner for nonsymmetric linear systems*, SIAM J. Sci. Comput., to appear.

[18] J. H. Bramble, *Multigrid Methods*, Longman Scientific and Technical, Harlow, U.K., 1993.

[19] A. Brandt, *Multilevel adaptive solutions to boundary value problems*, Math. Comp., 31 (1977), pp. 333–390.

[20] C. Brezinski, M. Redivo Zaglia, and H. Sadok, *Avoiding breakdown and near-breakdown in Lanczos type algorithms*, Numer. Algorithms, 1 (1991), pp. 199–206.

[21] W. L. Briggs, *A Multigrid Tutorial*, SIAM, Philadelphia, PA, 1987.

[22] P. N. Brown, *A theoretical comparison of the Arnoldi and GMRES algorithms*, SIAM J. Sci. Statist. Comput., 20 (1991), pp. 58–78.

[23] P. N. Brown and A. C. Hindmarsh, *Matrix-free methods for stiff systems of ODE's*, SIAM J. Numer. Anal., 23 (1986), pp. 610–638.

[24] T. F. Chan and H. C. Elman, *Fourier analysis of iterative methods for elliptic boundary value problems*, SIAM Rev., 31 (1989), pp. 20–49.

[25] R. Chandra, *Conjugate Gradient Methods for Partial Differential Equations*, Ph.D. dissertation, Yale University, New Haven, CT, 1978.

[26] P. Concus and G. H. Golub, *A generalized conjugate gradient method for nonsymmetric systems of linear equations*, in Computing Methods in Applied Sciences and Engineering, R. Glowinski and J. L. Lions, eds., Lecture Notes in Economics and Mathematical Systems 134, Springer-Verlag, Berlin, New York, 1976, pp. 56–65.

[27] P. Concus, G. H. Golub, and D. P. O'Leary, *A generalized conjugate gradient method for the numerical solution of elliptic partial differential equations*, in Sparse Matrix Computations, J. R. Bunch and D. J. Rose, eds., Academic Press, New York, 1976.

[28] J. Cullum, *Iterative methods for solving $Ax = b$, GMRES/FOM versus QMR/BiCG*, Adv. Comput. Math., 6 (1996), pp. 1–24.

[29] J. Cullum and A. Greenbaum, *Relations between Galerkin and norm-minimizing iterative methods for solving linear systems*, SIAM J. Matrix Anal. Appl., 17 (1996), pp. 223–247.

[30] J. Cullum and R. Willoughby, *Lanczos Algorithms for Large Symmetric Eigenvalue Computations, Vol. I. Theory*, Birkhäuser Boston, Cambridge, MA, 1985.

[31] J. Demmel, *The condition number of equivalence transformations that block diagonalize matrix pencils*, SIAM J. Numer. Anal., 20 (1983), pp. 599–610.

[32] J. J. Dongarra, J. R. Bunch, C. B. Moler, and G. W. Stewart, *LINPACK Users' Guide*, SIAM, Philadelphia, PA, 1979.

[33] J. Drkošová, A. Greenbaum, M. Rozložník, and Z. Strakoš, *Numerical stability of the GMRES method*, BIT, 3 (1995), pp. 309–330.

[34] V. Druskin, A. Greenbaum, and L. Knizhnerman, *Using nonorthogonal Lanczos vectors in the computation of matrix functions*, SIAM J. Sci. Comput., to appear.

[35] V. Druskin and L. Knizhnerman, *Error bounds in the simple Lanczos procedure for computing functions of symmetric matrices and eigenvalues*, Comput. Math. Math. Phys., 31 (1991), pp. 20–30.

[36] M. Dryja and O. B. Widlund, *Some domain decomposition algorithms for elliptic problems*, in Iterative Methods for Large Linear Systems, L. Hayes and D. Kincaid, eds., Academic Press, San Diego, CA, 1989, pp. 273–291.

[37] T. Dupont, R. P. Kendall, and H. H. Rachford, Jr., *An approximate factorization procedure for solving self-adjoint elliptic difference equations*, SIAM J. Numer. Anal., 5 (1968), pp. 559–573.

[38] M. Eiermann, *Fields of values and iterative methods*, Linear Algebra Appl., 180 (1993), pp. 167–197.

[39] M. Eiermann, *Fields of values and iterative methods*, talk presented at Oberwolfach meeting on Iterative Methods and Scientific Computing, Oberwolfach, Germany, April, 1997, to appear.

[40] S. Eisenstat, H. Elman, and M. Schultz, *Variational iterative methods for nonsymmetric systems of linear equations*, SIAM J. Numer. Anal., 20 (1983), pp. 345–357.

[41] S. Eisenstat, J. Lewis, and M. Schultz, *Optimal block diagonal scaling of block 2-cyclic matrices*, Linear Algebra Appl., 44 (1982), pp. 181–186.

[42] L. Elsner, *A note on optimal block scaling of matrices*, Numer. Math., 44 (1984), pp. 127–128.

[43] M. Engeli, T. Ginsburg, H. Rutishauser, and E. Stiefel, *Refined Iterative Methods for Computation of the Solution and the Eigenvalues of Self-adjoint Boundary Value Problems*, Birkhäuser-Verlag, Basel, Switzerland, 1959.

[44] V. Faber, W. Joubert, M. Knill, and T. Manteuffel, *Minimal residual method stronger than polynomial preconditioning*, SIAM J. Matrix Anal. Appl., 17 (1996), pp. 707–729.

[45] V. Faber and T. Manteuffel, *Necessary and sufficient conditions for the existence of a conjugate gradient method*, SIAM J. Numer. Anal., 21 (1984), pp. 352–362.

[46] V. Faber and T. Manteuffel, *Orthogonal error methods*, SIAM J. Numer. Anal., 24 (1987), pp. 170–187.

[47] K. Fan, *Note on M-matrices*, Quart. J. Math. Oxford Ser., 11 (1960), pp. 43–49.

[48] J. Favard, *Sur les polynomes de Tchebicheff*, C. R. Acas. Sci. Paris, 200 (1935), pp. 2052–2053.

[49] R. P. Fedorenko, *The speed of convergence of one iterative process*, U.S.S.R. Comput. Math. and Math. Phys., 1 (1961), pp. 1092–1096.

[50] B. Fischer, *Polynomial Based Iteration Methods for Symmetric Linear Systems*, Wiley-Teubner, Leipzig, 1996.

[51] R. Fletcher, *Conjugate gradient methods for indefinite systems*, in Proc. Dundee Biennial Conference on Numerical Analysis, G. A. Watson, ed., Springer-Verlag, Berlin, New York, 1975.

[52] G. E. Forsythe and E. G. Strauss, *On best conditioned matrices*, Proc. Amer. Math. Soc., 6 (1955), pp. 340–345.

[53] R. W. Freund, *A transpose-free quasi-minimal residual algorithm for non-Hermitian linear systems*, SIAM J. Sci. Comput., 14 (1993), pp. 470–482.

[54] R. W. Freund and N. M. Nachtigal, *QMR: A quasi-minimal residual method for non-Hermitian linear systems*, Numer. Math., 60 (1991), pp. 315–339.

[55] R. Freund and S. Ruscheweyh, *On a class of Chebyshev approximation problems which arise in connection with a conjugate gradient type method*, Numer. Math., 48 (1986), pp. 525–542.

[56] E. Giladi, G. H. Golub, and J. B. Keller, *Inner and outer iterations for the Chebyshev algorithm*, SCCM-95-10 (1995), Stanford University, Palo Alto. To appear in SIAM J. Numer. Anal.

[57] G. H. Golub and G. Meurant, *Matrices, moments, and quadratures* II *or how to compute the norm of the error in iterative methods*, BIT, to appear.

[58] G. H. Golub and D. P. O'Leary, *Some history of the conjugate gradient and Lanczos algorithms: 1948–1976*, SIAM Rev., 31 (1989), pp. 50–102.

[59] G. H. Golub and M. L. Overton, *The convergence of inexact Chebyshev and Richardson iterative methods for solving linear systems*, Numer. Math., 53 (1988), pp. 571–593.

[60] G. H. Golub and Z. Strakoš, *Estimates in Quadratic Formulas*, Numer. Algorithms, 8 (1994), pp. 241–268.

[61] G. H. Golub and R. S. Varga, *Chebyshev semi-iterative methods, successive overrelaxation iterative methods, and second-order Richardson iterative methods, parts* I *and* II, Numer. Math., 3 (1961), pp. 147–168.

[62] J. F. Grcar, *Analyses of the Lanczos Algorithm and of the Approximation Problem in Richardson's Method*, Ph.D. dissertation, University of Illinois, Urbana, IL, 1981.

[63] A. Greenbaum, *Comparison of splittings used with the conjugate gradient algorithm*, Numer. Math., 33 (1979), pp. 181–194.

[64] A. Greenbaum, *Analysis of a multigrid method as an iterative technique for solving linear systems*, SIAM J. Numer. Anal., 21 (1984), pp. 473–485.

[65] A. Greenbaum, *Behavior of slightly perturbed Lanczos and conjugate gradient recurrences*, Linear Algebra Appl., 113 (1989), pp. 7–63.

[66] A. Greenbaum, *Estimating the attainable accuracy of recursively computed residual methods*, SIAM J. Matrix Anal. Appl., to appear.

[67] A. Greenbaum, *On the role of the left starting vector in the two-sided Lanczos algorithm*, in Proc. Dundee Biennial Conference on Numerical Analysis, 1997, to appear.

[68] A. Greenbaum and L. Gurvits, *Max-min properties of matrix factor norms*, SIAM J. Sci. Comput., 15 (1994), pp. 348–358.

[69] A. Greenbaum, V. Ptak, and Z. Strakoš, *Any nonincreasing convergence curve is possible for GMRES*, SIAM J. Matrix Anal. Appl., 17 (1996), pp. 465–469.

[70] A. Greenbaum, M. Rozložník, and Z. Strakoš, *Numerical behavior of the MGS GMRES implementation*, BIT, to appear.

[71] A. Greenbaum and Z. Strakoš, *Predicting the behavior of finite precision Lanczos and conjugate gradient computations*, SIAM J. Matrix Anal. Appl., 13 (1992), pp. 121–137.

[72] A. Greenbaum and Z. Strakoš, *Matrices that generate the same Krylov residual spaces*, in Recent Advances in Iterative Methods, G. Golub, A. Greenbaum, and M. Luskin, eds., Springer-Verlag, Berlin, New York, 1994, pp. 95–118.

[73] L. Greengard and V. Rokhlin, *A Fast Algorithm for Particle Simulations*, J. Comput. Phys., 73 (1987), pp. 325–348.

[74] I. Gustafsson, *A class of 1st order factorization methods*, BIT, 18 (1978), pp. 142–156.

[75] M. H. Gutknecht, *Changing the norm in conjugate gradient-type algorithms*, SIAM J. Numer. Anal., 30 (1993), pp. 40–56.

[76] M. H. Gutknecht, *Solving linear systems with the Lanczos process*, Acta Numerica, 6 (1997), pp. 271–398.

[77] W. Hackbusch, *Iterative Solution of Large Sparse Systems of Equations*, Springer-Verlag, Berlin, New York, 1994.

[78] W. Hackbusch and U. Trottenberg, *Multigrid Methods*, Springer-Verlag, Berlin, New York, 1982.

[79] M. R. Hestenes and E. Stiefel, *Methods of conjugate gradients for solving linear*

systems, J. Res. Nat. Bur. Standards, 49 (1952), pp. 409–435.

[80] R. A. Horn and C. R. Johnson, *Matrix Analysis*, Cambridge University Press, London, U.K., 1985.

[81] R. A. Horn and C. R. Johnson, *Topics in Matrix Analysis*, Cambridge University Press, London, U.K., 1991.

[82] S. A. Hutchinson, J. N. Shadid, and R. S. Tuminaro, *Aztec User's Guide*, SAND95-1559, Sandia National Laboratories, Albuquerque, NM, 1995.

[83] A. Iserles, *A First Course in the Numerical Analysis of Differential Equations*, Cambridge University Press, London, U.K., 1996.

[84] D. Jespersen, *Multigrid methods for partial differential equations*, in Studies in Numerical Analysis, Studies in Mathematics 24, Mathematical Association of America, 1984.

[85] W. D. Joubert, *A robust GMRES-based adaptive polynomial preconditioning algorithm for nonsymmetric linear systems*, SIAM J. Sci. Comput., 15 (1994), pp. 427–439.

[86] W. D. Joubert and D. M. Young, *Necessary and sufficient conditions for the simplification of generalized conjugate-gradient algorithms*, Linear Algebra Appl., 88/89 (1987), pp. 449–485.

[87] D. Kershaw, *The incomplete Cholesky conjugate gradient method for the iterative solution of systems of linear equations*, J. Comput. Phys., 26 (1978), pp. 43–65.

[88] L. Yu. Kolotilina and A. Yu. Yeremin, *Factorized sparse approximate inverse preconditioning* I. *Theory*, SIAM J. Matrix Anal. Appl., 14 (1993), pp. 45–58.

[89] C. Lanczos, *An iteration method for the solution of the eigenvalue problem of linear differential and integral operators*, J. Res. Nat. Bur. Standards, 45 (1950), pp. 255–282.

[90] C. Lanczos, *Solutions of linear equations by minimized iterations*, J. Res. Nat. Bur. Standards, 49 (1952), pp. 33–53.

[91] E. W. Larsen, *Unconditionally Stable Diffusion-Synthetic Acceleration Methods for the Slab Geometry Discrete Ordinates Equations. Part* I: *Theory*, Nuclear Sci. Engrg., 82 (1982), pp. 47–63.

[92] D. R. McCoy and E. W. Larsen, *Unconditionally Stable Diffusion-Synthetic Acceleration Methods for the Slab Geometry Discrete Ordinates Equations. Part* II: *Numerical Results*, Nuclear Sci. Engrg., 82 (1982), pp. 64–70.

[93] E. E. Lewis and W. F. Miller, *Computational Methods of Neutron Transport*, John Wiley & Sons, New York, 1984.

[94] T. A. Manteuffel, *The Tchebychev iteration for nonsymmetric linear systems*, Numer. Math., 28 (1977), pp. 307–327.

[95] T. A. Manteuffel, *An incomplete factorization technique for positive definite linear systems*, Math. Comp., 34 (1980), pp. 473–497.

[96] T. Manteuffel, S. McCormick, J. Morel, S. Oliveira, and G. Yang, *A fast multigrid algorithm for isotropic transport problems* I: *Pure scattering*, SIAM J. Sci. Comput., 16 (1995), pp. 601–635.

[97] T. Manteuffel, S. McCormick, J. Morel, and G. Yang, *A fast multigrid algorithm for isotropic transport problems* II: *With absorption*, SIAM J. Sci. Comput., 17 (1996), pp. 1449–1475.

[98] S. McCormick, *Multigrid Methods*, SIAM, Philadelphia, PA, 1987.

[99] J. A. Meijerink and H. A. van der Vorst, *An iterative solution method for linear systems of which the coefficient matrix is a symmetric M-matrix*, Math. Comp., 31 (1977), pp. 148–162.

[100] N. Munksgaard, *Solving Sparse Symmetric Sets of Linear Equations by Precon-*

ditioned Conjugate Gradients, ACM Trans. Math. Software, 6 (1980), pp. 206–219.

[101] N. Nachtigal, *A look-ahead variant of the Lanczos algorithm and its application to the quasi-minimal residual method for non-Hermitian linear systems*, Ph.D. dissertation, Massachusetts Institute of Technology, Cambridge, MA, 1991.

[102] N. M. Nachtigal, S. Reddy, and L. N. Trefethen, *How fast are nonsymmetric matrix iterations?*, SIAM J. Matrix Anal. Appl., 13 (1992), pp. 778–795.

[103] N. M. Nachtigal, L. Reichel, and L. N. Trefethen, *A hybrid GMRES algorithm for nonsymmetric linear systems*, SIAM J. Matrix Anal. Appl., 13 (1992), pp. 796–825.

[104] R. A. Nicolaides, *On the L^2 convergence of an algorithm for solving finite element equations*, Math. Comp., 31 (1977), pp. 892–906.

[105] A. A. Nikishin and A. Yu. Yeremin, *Variable block CG algorithms for solving large sparse symmetric positive definite linear systems on parallel computers*, I: *General iterative scheme*, SIAM J. Matrix Anal. Appl., 16 (1995), pp. 1135–1153.

[106] Y. Notay, *Upper eigenvalue bounds and related modified incomplete factorization strategies*, in Iterative Methods in Linear Algebra, R. Beauwens and P. de Groen, eds., North–Holland, Amsterdam, 1991, pp. 551–562.

[107] D. P. O'Leary, *The block conjugate gradient algorithm and related methods*, Linear Algebra Appl., 29 (1980), pp. 293–322.

[108] C. W. Oosterlee and T. Washio, *An evaluation of parallel multigrid as a solver and a preconditioner for singular perturbed problems*, Part I. *The standard grid sequence*, SIAM J. Sci. Comput., to appear.

[109] C. C. Paige, *Error Analysis of the Lanczos Algorithm for Tridiagonalizing a Symmetric Matrix*, J. Inst. Math. Appl., 18 (1976), pp. 341–349.

[110] C. C. Paige, *Accuracy and Effectiveness of the Lanczos Algorithm for the Symmetric Eigenproblem*, Linear Algebra Appl., 34 (1980), pp. 235–258.

[111] C. C. Paige and M. A. Saunders, *Solution of sparse indefinite systems of linear equations*, SIAM J. Numer. Anal., 11 (1974), pp. 197–209.

[112] B. N. Parlett, *The Symmetric Eigenvalue Problem*, Prentice–Hall, Englewood Cliffs, NJ, 1980.

[113] B. N. Parlett, D. R. Taylor, and Z. A. Liu, *A look-ahead Lanczos algorithm for unsymmetric matrices*, Math. Comp., 44 (1985), pp. 105–124.

[114] C. Pearcy, *An elementary proof of the power inequality for the numerical radius*, Michigan Math. J., 13 (1966), pp. 289–291.

[115] J. K. Reid, *On the method of conjugate gradients for the solution of large sparse linear systems*, in Large Sparse Sets of Linear Equations, J. K. Reid, ed., Academic Press, New York, 1971.

[116] Y. Saad, *Preconditioning techniques for nonsymmetric and indefinite linear systems*, J. Comput. Appl. Math., 24 (1988), pp. 89–105.

[117] Y. Saad, *Iterative Methods for Sparse Linear Systems*, PWS Pub. Co., Boston, MA, 1996.

[118] Y. Saad and A. Malevsky, *PSPARSLIB: A portable library of distributed memory sparse iterative solvers*, in Proc. Parallel Computing Technologies (PaCT-95), 3rd International Conference, V. E. Malyshkin, et al., ed., St. Petersburg, 1995.

[119] Y. Saad and M. H. Schultz, *GMRES: A generalized minimal residual algorithm for solving nonsymmetric linear systems*, SIAM J. Sci. Statist. Comput., 7 (1986), pp. 856–869.

[120] H. A. Schwarz, *Gesammelte Mathematische Abhandlungen*, Vol. 2, Springer, Berlin, 1890, pp. 133–143 (first published in Vierteljahrsschrift Naturforsch. Ges. Zurich, 15 (1870), pp. 272–286).

[121] H. D. Simon, *The Lanczos algorithm with partial reorthogonalization*, Math.

Comp., 42 (1984), pp. 115–136.

[122] G. L. G. Sleijpen, H. A. Van der Vorst, and J. Modersitzki, *The Main Effects of Rounding Errors in Krylov Solvers for Symmetric Linear Systems*, Preprint 1006, Universiteit Utrecht, The Netherlands, 1997.

[123] B. Smith, P. Bjorstad, and W. Gropp, *Domain Decomposition. Parallel Multilevel Methods for Elliptic Partial Differential Equations*, Cambridge University Press, London, U.K., 1996.

[124] P. Sonneveld, *CGS, a fast Lanczos-type solver for nonsymmetric linear systems*, SIAM J. Sci. Statist. Comput., 10 (1989), pp. 36–52.

[125] G. Strang and G. J. Fix, *An Analysis of the Finite Element Method*, Prentice-Hall, Englewood Cliffs, NJ, 1973.

[126] Z. Strakoš, *On the real convergence rate of the conjugate gradient method*, Linear Algebra Appl., 154/156 (1991), pp. 535–549.

[127] K. C. Toh, *GMRES vs. ideal GMRES*, SIAM J. Matrix Anal. Appl., 18 (1997), pp. 30–36.

[128] C. H. Tong, *A Comparative Study of Preconditioned Lanczos Methods for Nnsymmetric Linear Systems*, Sandia report SAND91-8240, 1992.

[129] L. N. Trefethen, *Approximation theory and numerical linear algebra*, in Algorithms for Approximation II, J. Mason and M. Cox, eds., Chapman and Hall, London, U.K., 1990.

[130] R. Underwood, *An Iterative Block Lanczos Method for the Solution of Large Sparse Symmetric Eigenproblems*, Technical report STAN-CS-75-496, Computer Science Department, Stanford University, Stanford, CA, 1975.

[131] A. van der Sluis, *Condition numbers and equilibration matrices*, Numer. Math., 14 (1969), pp. 14–23.

[132] A. van der Sluis and H. A. van der Vorst, *The rate of convergence of conjugate gradients*, Numer. Math., 48 (1986), pp. 543–560.

[133] H. A. van der Vorst, *The convergence behavior of preconditioned CG and CG-S in the presence of rounding errors*, in Preconditioned Conjugate Gradient Methods, O. Axelsson and L. Kolotilina, eds., Lecture Notes in Mathematics 1457, Springer-Verlag, Berlin, New York, 1990.

[134] H. A. van der Vorst, *Bi-CGSTAB: A fast and smoothly converging variant of Bi-CG for the solution of nonsymmetric linear systems*, SIAM J. Sci. Comput., 13 (1992), pp. 631–644.

[135] R. S. Varga, *Matrix Iterative Analysis*, Prentice-Hall, Englewood Cliffs, NJ, 1962.

[136] R. S. Varga, *Factorization and normalized iterative methods*, in Boundary Problems in Differential Equations, R. E. Langer, ed., 1960, pp. 121–142.

[137] P. Vinsome, *Orthomin, an iterative method for solving sparse sets of simultaneous linear equations*, in Proc. 4th Symposium on Numerical Simulation of Reservoir Performance, Society of Petroleum Engineers, 1976, pp. 149–159.

[138] V. V. Voevodin, *The problem of a non-selfadjoint generalization of the conjugate gradient method has been closed*, U.S.S.R. Comput. Math. and Math. Phys., 23 (1983), pp. 143–144.

[139] H. F. Walker, *Implementation of the GMRES method using Householder transformations*, SIAM J. Sci. Statist. Comput., 9 (1988), pp. 152–163.

[140] R. Weiss, *Convergence Behavior of Generalized Conjugate Gradient Methods*, Ph.D. dissertation, University of Karlsruhe, Karlsruhe, Germany, 1990.

[141] P. Wesseling, *An Introduction to Multigrid Methods*, Wiley, Chichester, U.K., 1992.

[142] O. Widlund, *A Lanczos method for a class of nonsymmetric systems of linear equations*, SIAM J. Numer. Anal., 15 (1978), pp. 801–812.

[143] H. Wozniakowski, *Roundoff error analysis of a new class of conjugate gradient algorithms*, Linear Algebra Appl., 29 (1980), pp. 507–529.

[144] D. M. Young, *Iterative Solution of Large Linear Systems*, Academic Press, New York, 1971.

[145] D. M. Young and K. C. Jea, *Generalized conjugate gradient acceleration of nonsymmetrizable iterative methods*, Linear Algebra Appl. 34, 1980, pp. 159–194.

[146] H. Yserentant, *Old and new convergence proofs for multigrid methods*, Acta Numerica, 2 (1993), pp. 285–326.

Index